SANKOFA?

*How Racism and Sexism
Skewed New York's
Epochal Black Research Project*

ALSO BY DAVID R. ZIMMERMAN

Rh, The Intimate History of a Disease and Its Conquest

To Save a Bird in Peril

The Essential Guide to Nonprescription Drugs

Zimmerman's Complete Guide to Nonprescription Drugs

The Doctor's Anti-Breast Cancer Diet
(with Sherwood L. Gorbach, M.D.)

SANKOFA?

*How Racism and Sexism
Skewed New York's
Epochal Black Research Project*

David R. Zimmerman
with Michelle Mitchell

Probe Press, Sheffield, VT

First Edition

Copyright © 2013 by David R. Zimmerman

All rights reserved. No part of this book may be reproduced or transmitted in any form or by any means electronic or mechanical, including by photocopying, by recording, or by any information storage and retrieval system, without the express permission of the author and publisher, except where permitted by law.

Published in the United States of America
by Probe Press, P.O. Box 78, Sheffield, VT 05866.
Book available at Amazon.com, CreateSpace E-store
(www.createspace.com/3792046), and other retail stores.

ISBN 978-1-477568-35-4
Library of Congress Catalog No. 2012909866

*Book Design by Barry Sheinkopf for Bookshapers
(www.bookshapers.com)*

For Veva forever . . . .

"Negroes must be honest enough to admit that our standards do often fall short. One of the sure signs of maturity is the ability to rise to the point of self-criticism. Whenever we are objects of criticism from white men . . . we must pick out the elements of truth, and make them the basis of creative reconstruction."
— *Martin Luther King, 1958*

"We must . . . stop tolerating the . . . distortion of science."
— *Al Gore, 2007*

TABLE OF CONTENTS

Prologue, *1*
Foreword, *5*
Introduction, *7*
Chapter 1: Grave Findings, *17*
Chapter 2: Procedures, *28*
Chapter 3: A Defining Moment, *39*
Chapter 4: Going Forward, *51*
Chapter 5: Taking Over, *63*
Chapter 6: Counterattack, *75*
Chapter 7: Science and Inspiration, *87*
Chapter 8: Moving Bodies, *98*
Chapter 9: What Might Be Found, *106*
Chapter 10: Digging In, *116*
Chapter 11: A Musket Ball and Beads, *124*
Chapter 12: Studying the Bones, *134*
Chapter 13: Loggerheads, *142*
Chapter 14: Racism, *152*
Chapter 15: War, *158*
Chapter 16: Data and Discord, *168*
Chapter 17: Rejection, *178*
Chapter 18: Genetic Forebears, *186*
Chapter 19: Secrets, *198*
Chapter 20: GSA Takes Over, *205*
Chapter 21: Betrayal, *210*
Chapter 22: 'Community' Views, *216*
Chapter 23: Finish Line!, *224*
Chapter 24: What They Found, *233*
Chapter 25: Is It Science?, *242*
Chapter 26: What Went Wrong?, *248*
Chapter 27: Findings, *252*
Epilogue, *257*
About the Author, *263*
Sources, *265*
Bibliography, *281*
People In This Book, *283*
Time Line, *290*
Indices, *299*

SANKOFA?

PROLOGUE

MOST BOOKS ABOUT SCIENCE are success stories: research advances and challenges.

This book isn't.

It's an exposé of a major failure: the federally financed study of 400 sets of human remains — presumed to be black human remains — dug up two decades ago from a cemetery near the City Hall in New York City. This site is called the African Burial Ground; black men, women, and children who died during the 18th century were laid to rest there. The area was covered over with rubble and then buildings, and the cemetery was mostly forgotten — until the federal government decided to erect a huge office building on the site. In 1991, federal contractors began tearing down and digging out the existing buildings in order to make a basement with an underground parking garage. The cemetery — human bones and remnants of coffins — soon came into view.

This was an archaeological treasure, and especially for black people, it was a profound emotional treasure as well: a human link between their lives and their African forebears. Studying these moldering bones might tell much about those Afro-Americans and the lives they lived in bondage. This treasure, unfortunately, has been squandered. Very few scientific results were published before the bones were returned to the earth a few years ago.

Some 400 sets of remains were recovered (leaving hundreds or thousands more behind). Over the protest of local scientists — the archeologists, anthropologists, and forensic pathologists who had disinterred them — government officials sent the remains to a young anthropologist in Washington, D.C. The reason for the change, this anthropologist had insisted, was that the New Yorkers were white, while he was black. And only a black person was qualified to study and understand these ancestral remains. White people were not.

Rancor grew between the black scientist, Michael L. Blakey, PhD, of Howard University, and the white New Yorkers of the Metropolitan Forensic Anthropology Team. They were white racists, Blakey charged. He was imperious, a weak but aggressive scientist, the team and their colleagues retorted.

In a contract between Howard University and the federal government, Blakey promised to report all that could be learned from the remains in about five years, at a cost that quickly grew to over $5 million. He missed his deadline and then, with extensions, another and another. His massive Final Report was not delivered until 2004. The data have never been published in a peer-reviewed scientific journal. So, despite all the cost and effort, the findings do not qualify as science, in my view, as explained below. Since the bones were reburied, the study can't be redone.

Scientists and science are the losers. So is the black "Descendant Community," who had grudgingly allowed the work to proceed on the premise that scientific information would justify the disturbance and damage to the bones. All Americans — black and white — have lost, since these grim findings now fall outside the annals of science.

SANKOFA?

In short, the study turned out to have been a fiasco in my view. But this is not widely known. Blakey and the General Services Administration (GSA), the sponsoring federal agency, have hidden it now for over a decade, starting with the agency's rejection of Blakey's first draft of the report as unacceptable. Shortly thereafter he was quietly removed from leadership of the study.

I find it improbable that these failures could go unremarked for over a decade. That the federal government would sponsor a racially based and biased "scientific" investigation for over a decade, at a cost of millions. And that no one complained. Or said anything about it. But that is what happened.

Can a misdirected scientific endeavor originating in the heart of New York City, and in the hearts of a vocal minority of citizens, go unchallenged for 10 — and now over 20 — years? Sadly, the evidence shows that it can.

Belatedly, that evidence is now here, in the pages that follow. I will describe the false starts and will trace them . . . to failure. I will show how this treasure was squandered. How. And by whom.

David Zimmerman

Sankofa?

FOREWORD

THIS LARGE STONE ORNAMENT was cut into the memorial for denizens of the African Burial Ground (ABG) in Lower Manhattan. It is a rendering of a putative *Sankofa*, a West African symbol made of 18th-century iron tacks (nails) that was found on a man's coffin lid unearthed at the site (Burial #101).

This design has been identified as a Sankofa by anthropologist Michael L. Blakey, PhD, who directed the study of the cemetery's human remains. He said it is the only graphic notation discovered among the 400 sets of human remains disinterred there in the last century. He has called it "perhaps the most ethno-linguistically specific material cultural evidence recovered from the site."

Other researchers have been less certain. The heart shape led some to suggest that it means, simply, "I ♥ you!" — a last farewell from the bereaved.

Blakey assigns this "Sankofa" a cultural meaning rather than a personal one: In his view, the brass tacks depict two birds, face-to-face, each with its head turned backwards. He says that this symbolizes a West African (Adinkra) proverb: "It is not taboo to return and fetch it when you forget." In plain English, this means: We look to the past in order to understand the present and build the future.

Blakey and his co-workers adopted this motif as an expression of black Americans' commitment to the African Burial Ground and of their own commitment to study and honor its inhabitants. Black supporters of the research and the federal government (General Services Administration), which funded it, refined the symbol to this:

The symbolic meaning of Burial #101's "Sankofa" may not yet be resolved. But the wisdom of studying the past to understand the present and plan for the future is unarguably valid.

In this report, the past that needs to be probed is not only the horrors of black enslavement in 17th- and 18th-century New York City. It also encompasses the efforts of our contemporaries who dug up and set out to interpret these remains for black people and all other Americans. In so doing, have they honored, with truth, these "sacred" bones?

SANKOFA?

INTRODUCTION

AS A MANHATTAN RESIDENT and as a science writer, I, like very many New Yorkers, was captivated by the rediscovery of the African Burial Ground (ABG) in 1991 — which is now quite a while ago. The graves and the skeletal remains in this cemetery promised to be a grim but revealing window through which to look back, before modern times, to life and human bondage during New York's colonial years under Dutch and English rule.

The cemetery was used for burials between about 1710 and 1795, historical studies showed. Most, but by no means all who had been laid to rest there were black people — men, women, and children, including babes in their mothers' arms. Most had died as slaves — though there was not then, in 1991, and still is not now any way to discern who among these people were enslaved when they died and who died free.

They were rediscovered by government contractors digging a basement parking garage in Lower Manhattan. As the shovels turned and more and more human remains emerged, bitter conflict emerged: One party to the conflict was the General Services Administration (GSA), the US government's construction and maintenance agency, which, under tight deadlines, was building a giant, 34-story federal office building, 290 Broadway, on the site; it was due to open before century's end. Another party to the conflict was a loosely knit coalition of local anthropologists, archeologistss, and other preservationists who sought to save what remained of the cemetery. They wanted to study the human remains and any related artifacts that could be snatched from harm's way. There was conflict, finally, between GSA and the preservationists on the one hand and some members of New York's large black community on the other.

The blacks — the *Descendant Community* as they came to call themselves — were not just intrigued by the discoveries. They were also outraged. "We consider this sacred ground," declared Bill Davis, a black architect and member of the city's Landmarks Preservation Commission. The descendants were infuriated by the desecration of their antecedents' resting place and remains. They were angered by GSA's refusal to hear and act on their concerns. The fed's cavalier unconcern for the descendants' forebears resonated with the 20th-century racism that blighted many of their own lives, stoking their antipathy to white people and white government. Rage and the threat of violence hung in the air.

The conflict shifted when a black anthropologist, Michael L. Blakey, PhD, a credentialed scholar, appeared on the scene. He directed a scientific facility, the W. Montague Cobb Biological Anthropology Laboratory at Howard University, an historical black university in Washington, D.C.*

Blakey rallied the black community. He attacked federal officials at GSA as uncaring. He denounced the white re-

---

* Cobb was a physician and physical anthropologist at Howard's medical school in the mid-20th-century. He earned a well-deserved reputation for fighting white racist ideas, particularly the attempt to correlate black Americans' head shapes and facial features with various types of crime. Cobb was the first African-American to earn a PhD in anthropology.

searchers who were disinterring the bones as incompetent. He strongly suggested that only black scientists were qualified to handle, study, honor, and understand these historic black remains — and he was the right man for the job. What is more, he would rescue the remains from ill handling in New York by having them shipped to Howard University for study. There would be money enough to support his research agenda and to rebury and memorialize these forebears, as provided by the *National Historic Preservation Act of 1966* (NHPA). This federal law had been written to meet just such situations: an historic site threatened by new government or governmentally sponsored construction. Up to two percent of the new project's cost should be available from the Federal government for rescue or rehabilitation or for mitigating damage or unavoidable destruction of an historical site. The budget for the new office building, which was to house offices for the Federal courts in Manhattan, had reached $276 million by the early 1990s.

I DID NOT WRITE about the ABG at the time. But I was keenly interested in it — and I became keenly disturbed by the wild rhetoric it engendered.

The irrationality of such rhetoric was of particular concern to me because my journalistic focus in the 1990s was the relationship — and very often the collisions — between science (including medicine) on the one hand and irrational value systems on the other. Science *versus* religion. Rationality *versus* spiritualism. Knowledge *versus* belief.

In a larger context then, this account is part of the great public debate: Science *versus* Faith. While it does not involve organized religion, the conflict between the mostly white scientists who initiated the ABG work and the mostly black scientists who set out to complete it is a struggle between objective research and research that starts out with an ideological — and in their words *vindicative* — point of view.

Science's enemies in this context have included: Pro-life, Animal Rights, alternative medicine, and Creationism (now renamed Intelligent Design). All hot topics, then and now. In each case, the True Believers denigrated individual free choice in favor of faith-based movements and authoritarian leadership. These were — are — the opposite of individual informed choice, which I thought was (and hope is still) the essential basis, the underpinning, of democracy. Since science's goal is to inform, the invariably dishonest or distorted ideologies arrayed against it also threaten democracy and individual freedom.

I took sides. And I reported the findings of my journalistic investigations in the newsletter that I published, called PROBE. I produced 99 issues before stopping, in the year 2000. This report is a continuation of that investigative science journalism; I consider the book that you are holding to be PROBE #100. My final PROBE.

Back then, reading about Michael Blakey's continuing rant, I heard anti-science, or, at best, belief over science. Yet Blakey was a scientist, a doctoral recipient from a prestigious school, the University of Massachusetts.

Here, I said then, is an irreconcilable conflict. I thought that it would be a disaster. And so, more than two decades later, it seems clear that very little science has been done, and at great cost. But when Blakey was quietly relieved of his leadership role, in 2002, virtually no one was aware of what had happened. One reason is that a voluminous flow of promotional information — "education and interpretation" — paid for by the feds at a cost of millions of dollars, never disclosed

SANKOFA?

the paucity of the scientific findings.*

The GSA appears never to have taken stock of its uninvited and unsettling encounter with the ABG and its modern-day champions: The public affairs officer at its New York regional headquarters said, in 2007, "I know of no after [action] report on this that has ever been prepared." By the same token, it appears that no GSA official has ever publicly shared fully his or her — or the agency's — trials and tribulations in this endeavor.

The GSA official who worked most closely with Michael Blakey was an engineer and urban planner named Peter Sneed. "[Blakey] was a difficult person to deal with," recalls Sneed, who in fact dealt with him longer and more directly, than any other agency official. "He stormed into GSA with guns blazing! Every problem he ever had was our fault."

On the surface and in public statements, there have been few indications of this discord, or of the failures that resulted. The partners have shut out the press and apparently all other inquirers. Of the thousands, perhaps tens of thousands, of news stories, articles, webcasts, telecasts, and broadcasts that have appeared about the ABG in the last two decades, only a very few have scratched the surface and exposed the enormous problems beneath. The major media, including the New York Times, simply have not bothered to look critically at what happened.

So, can a scientist with strong personal feelings about a subject study it as a researcher? He or she can. But it's hard. And the stronger one's feelings, the harder it is:

Science and personal belief — religion or politics, for ex-

---

* GSA paid for much of this work through a contract with a West Chester, Pennsylvania, company, John Milner Associates.

ample — are mutually exclusive enterprises. They can't be melded together (albeit lots of people try). They don't meet in the middle; there is no interface. Rudyard Kipling said it succinctly a century ago: "East is East and West is West, and n'er the twain shall meet."

"East" here — what used to be called *The Orient* — is mysterious and intuitive and spiritual. The "West" — us — is fact-based, materialistic, and rational.

How can a religious person, a believer, do science? Such a person can, to the extent that he or she is able to suspend his or her belief on the subject at hand. But it is much harder to do objective science if one has strong subjective views on a matter. *Particularly*, if one co-opts the science to support, prove, rebut, or deny an item or items of faith. The harder one tries to bend or manipulate science to support religious or political beliefs, the weaker and more erroneous the results become.

Is a True Believer thus unable to study objectively the foundations of his or her beliefs? Never say never. But it is particularly difficult — and requires an exceptional ability to suspend these key beliefs while the work is in progress.

Michael Blakey injected himself into the ABG scientific project based on what he claimed were deeply felt beliefs about his ancestors and their meaning for contemporary life. He thus needed to proceed with extraordinary objectivity and care to justify the key role he planned for himself in the project. Could he do it? Would he do it? I will explore these important questions.

BY 2001, WHEN THE contractually required scientific reports still had not appeared, it became clear to me that something was seriously wrong. Initial probing, against great resistance, confirmed this. GSA refused to provide me substantive information. The ABG's educational office also refused all substantive comment and referred questions — such as, What has happened? — to Blakey. But Blakey, who by then had left Howard for a prestigious new job at the College of William & Mary, in Williamsburg, Virginia, did not return telephone calls or respond to mail or e-mail inquiries. I was stone-walled.

Stone walls suggest that something is hidden inside. Stone walls confront — and challenge — reporters.

We are admonished by editors to "Get it first!" And, like scientists, we also have the vocational imperative: "Get it right!" This is what I have set out to do here.

There is one additional reason why reporters, or others, may have shied away from seeking an in-depth look at the ABG project: fear of being tarred as "racists," particularly "white racists." This has been Blakey's standard rejoinder to the federal government and its representatives — the hands that fed him these many years. This is his response to critics. "Racist" may well be his reaction to this inquiry. In his mind, I may be one.

I am white. But it is fair to point out that my family — my late wife and our children — are black.

I'D LIKE TO MAKE clear at the start what this book is about and what it is not. It is *not primarily* about the history of the ABG or the lives of black men and women in Dutch or English Manhattan. These topics are enormous challenges for historians but are outside my purview. Neither is this a considered report on the scientific findings of the ABG researchers; I am not a scientist.

Rather, this is a report on what went wrong — a subject that appears not to have previously been independently examined. It is, or at least should be, of major interest to the ABG constituencies, or *stakeholders* as they are called in the jargon of public politics. These stakeholders include the Descendant Community. They also include New Yorkers of all ethnic backgrounds. The ABG was — is — a major discovery and a major historical event in their city. New Yorkers are curious and they also are prideful. This is their historical discovery, even though the bodies and hence the research focus were shifted to Washington before being returned to the ABG site for reburial. All Americans are stakeholders, moreover, because the presence of slaves and slaves' graves in Manhattan is a poorly known but essential part of our history.

One stipulation: Black people, enslaved or free, were treated badly and lived poorly in colonial New York City.* How badly and how poorly is still being teased out by the ABG project and other research. I won't gainsay the worst of these findings. But more, and more-particular, data are needed. It is horrendous that black men and women — alleged plotters and revolutionaries — were burned at the stake by white colonialists near the ABG in the mid-18th century.** But was this the hideous norm, or a rare extreme? Is the torture and killing of Iraqi prisoners a reflection of ordinary Americans' morality in the 21st century or, rather, a brutal aberration? These are difficult questions — and ones I will not try to answer here.

---

* But not all blacks were anonymous and down-trodden back then. One is known to every New York City student who has studied American history. He is Samuel Fraunces, also known as "Black Sam." A free, black West Indian, he established Fraunces Tavern in 1762, according to the Schomburg Center for Research in Black Culture. George Washington dined there during the Revolutionary War. Fraunces himself managed a later banquet there, in 1783, where Washington said farewell to his commanders after they beat the British. When Washington became president, he appointed Fraunces as steward of his executive mansion in New York City. Fraunces Tavern continues in business on Pearl Street in Manhattan.

** Harvard historian Jill Lepore has documented much horrendous data in her *New York Burning* (Knopf, 2005).

SANKOFA?

A word must be said about who helped — and who didn't — on this journalistic project. Interviewees are identified in the main text. Major contributors are cited — and thanked! — in the Acknowledgments. But a number of key participants in the ABG project, whose reminiscences could have added clarity and detail, declined to help.

Project director Michael Blakey did not respond to any of my many requests — by phone, letter, and e-mail — for interviews. His second-in-command, Lesley Rankin-Hill, PhD, answered a single phone query and then stopped. His lab director, Mark Mack, volunteered nothing of note. Neither did Cheryl LaRoche, PhD, of John Milner Associates, who has a different professional perspective that I was eager to hear. Educator Sherill D. Wilson, PhD, ducked tough questions.

I made numerous attempts to interview Michael Blakey while I was investigating and writing this book. He was unresponsive. I made a final effort to elicit his views when the manuscript was nearing completion. I mailed and then e-mailed him the text, and invited specific comment to any and all assertions it contained for me to consider as I finalized the manuscript, and also offered to add an Addendum, written by him, that would be independently edited. I promised him a line on the book cover saying "Rejoinder by Dr. Michael Blakey."

Prior to reading the text, Blakey condemned this book, saying it "has seemed ill- motivated from the beginning." He said, subsequently, that it showed my "deep delusion and pitiful hatred" of him and the African Burial Ground Project. He declined my offer of a published rejoinder. Nor did he provide any specific challenge to any particular assertions contained in this book.

At GSA, contracts and Freedom of Information officer Mildred Broughton, whom her boss, Alan Greenberg calls "the

David Zimmerman

work horse" of the ABG Project, released as little information as possible over a half-dozen years. Other than Peter Sneed, no GSA official was willing to go on record with me to evaluate his or her relationship with researcher Blakey. These participants' lack of candor and of clarity have been monumental.

THINGS WORKED THIS WAY: The NHPA must be invoked when a government agency seeks to destroy, damage, or disrupt a US historical site. Under the law, efforts may be made to stop or change the new development. If that is not possible, efforts must be made to mitigate the loss by salvaging some or all of the historical remains and curating them as a monument or in a museum — by creating and sustaining an educational program that will keep them in public view.

What will this education teach? If it is truthful, it can only teach what has been learned through research on the skeletal remains and related artifacts and other information. If the research is accurate, and if it is accurately presented to educators, scholars, and the public, then the lessons can be truthful. But if the research findings are inadequate, or distorted, the lessons cannot speak truth.

For this reason, educators, now and in the future, have a high stake in the accuracy and integrity of the research work. Are the many claims that are being based on the data accurate and thus worthy of belief? Or are they distortions that should be discounted?

I hope that this report will begin to answer these questions. I'll welcome questions and comments at *sankofa-book.com*.

—David R. Zimmerman
Sheffield, Vermont
January 2013

SANKOFA?

1

GRAVE FINDINGS

THE ORIGINS AND EARLY history of the African Burial Ground (ABG) — originally called the Negroes Burial Ground — are little known.

The site, today, is underneath and adjacent to a 34-story federal office building at 290 Broadway, just north of City Hall, in Manhattan. The Federal Bureau of Investigation is one of the tenants.

Congress named the building the Ted Weiss Federal Building, in memory of a liberal, white, Democratic congressman from Manhattan who had a reputation for being the conscience of Congress. Black leaders objected to the name, unsuccessfully, saying that the building should memorialize the black people buried beside and beneath it.

In the late 17th century, this site was part of the Commons, a rustic area north of the Palisades, the high wooden fence built to protect the city from attacks. Evidence has been found showing that people were buried there as early as 1712 and possibly earlier. In their book *Unearthing Gotham* (Yale, 2001), archeologists Anne-Marie Cantwell, PhD, of Rutgers University, in New Brunswick, New Jersey, and Diana diZerega Wall, PhD, of the City College of New York, write that in 1697 the City Fathers segregated burials racially by ordering that "blacks, Jews, and Catholics" could not be buried in the Trinity Church cemetery farther downtown, which previously had been the city's potter's field. Blacks chose or were assigned the site on the Commons, which also was used for grazing cattle and other public purposes, including executions.

Africans had first been brought to New York (New Amsterdam it then was called) as slaves, in 1626, right after the colony was founded.* Slavery continued, more harshly, after the English took control, in 1664.

By century's end, there were 700 blacks in New York City —14 percent of the population. Many, probably most, were slaves. Some were free men and women. The British imported and used slaves in their farms, businesses, and homes. Many were brutally treated.

Blacks continued to bury their dead in the Negroes Burial Ground until about 1795, when it was closed. A new cemetery designated for blacks was then opened further uptown.

No contemporary description of the Negroes Burial Ground has come to light. Many years later, in 1865, a New York City clerk wrote:

> *Beyond the Commons lay what in the earliest settlement of the town had been appropriated as a burial place for negroes [sic], slaves and free. It was a desolate, unappropriated spot, descending*

---

*The first resident, before the Dutch, was a black man from Hispaniola. His mother was an African slave. Jan Rodriguez spent the winter of 1613–1614 on what is now Governor's island off of Wall Street, and may have stayed much longer. He trapped animals and traded for pelts with the local Native Americans.

SANKOFA?

with a gentle declivity toward a ravine which led to the Kalkhook [Collect, fresh-water] pond. The negroes in this city were, both in the Dutch and English colonial times, a proscribed and detested race, having nothing in common with the whites. Many of them were native Africans, imported hither in slave ships, and retaining their native superstitions and burial customs, among which was that of burying by night, with various mummeries and outcries.

A less dire view of the cemetery and how it came into being outside the Palisades has recently been advanced by ABG historians: The Dutch needed food crops from farming, and they also sought buffers from attacks from the north by Englishmen from Massachusetts and what is now Connecticut. So, enslaved and free blacks were granted land to start farms there. The cemetery was set up in their midst. They did not, however, own title to the cemetery land. A white family did.*

Because of the isolation, and because of the gravity of the funerals, recent researchers assumed that the graves might reveal information on burial customs as well as other traits retained from the buried persons' African lives and forebears. For example, following African tradition, it has been found that the bodies were buried with their heads to the west, so that when they sit up, on Judgment Day, they will face east toward the rising sun. But more specific comparisons would prove to be difficult, since funerary practices in the African areas where slaves were captured were little known to outsiders — and thus were poorly recorded.

Given the wide diversity of African and American people laid to rest there, researchers were surprised by the uniformity

---

* The adjacent property was owned by Abraham Pieterszen, who arrived in 1625. He was the patriarch of the Van Dusen family, who were slave owners. Pieterszen's progeny — 200,000 descendants — include President Martin Van Buren and President Franklin Roosevelt.

of their burials; most lay face up in rectangular or tapered coffins. This uniformity might suggest that the blacks had a protocol for proper burial. More likely, however, the protocol was for *white* burials — for which, no doubt, blacks dug most of the graves.

Not all the buried black people were slaves. Some were freemen and women. What is more, not all the bodies were black. Cantwell and Wall, citing earlier sources, say that during the American Revolution, the English, who occupied Manhattan, buried black and white American prisoners, many of whom they executed secretly, in the Negroes Burial Ground.

At least two different methods for distinguishing black peoples' remains from white peoples' have been discussed with regard to the ABG. But few data have as yet been published on how many of the 400 disinterred remains were not black people.

By the late 18th century, New York City's growth northward on Manhattan Island was reaching beyond the Burial Ground. The Commons area had become the city's civic and political center — as it is to this day.

The cemetery was closed. The land was surveyed and divided into lots for residences, commerce, and manufacturing. As the 19th century progressed, larger, heavier buildings were built on the site for factories, warehouses, and offices. Many of the buildings had sub-basements that went down 20 or more feet below ground level. Little if any thought was given to the bodies that lay there.*

A CENTURY LATER, IN 1989, the General Services Administration (GSA), the real estate, construction, and building main-

---

* The actual number, whether one is counting grave shafts or sets of human remains, has varied through the years from 390 to more than 425. For continuity, the number 400 will be used here, unless otherwise noted for statistical purposes.

Sankofa?

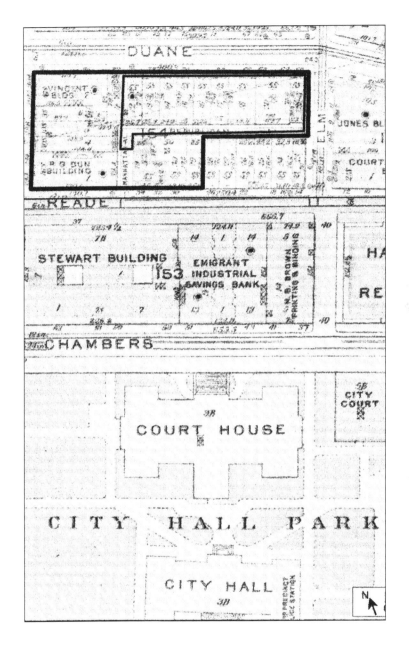

At Civic Center. African Burial Ground (inner line) and National Historic Landmark (outer line) are superimposed on 1980s Manhattan map in image from ABG *Archeology Final Report*. Federal building now towers over site.

tenance agency of the United States government, began work on a much-needed office building that was to cover part of the cemetery site. The GSA is a major part of the federal government: It manages over one quarter of the fed's procurement dollars and influences the management of $500 billion in US assets. Among the 8,300 buildings within its purview, the agency says that it is "steward" to more than 420 historic properties. (The agency is not immune to excess and corruption. In 2010, West Coast GSA officials put on a lavish conference at a posh resort and casino near Las Vegas, and flew in 300 guests. They played for four days. The cost: over $800,000 of taxpayer money, as congressmen angrily reminded the GSA's leader — who was forced to resign.)

The GSA was required, under the National Historic Preservation Act of 1966, to define the new construction's impact on the natural and cultural features of the site. GSA hired an engineering company, which in turn hired a small, New Jersey-based archaeological consulting company, Historic Conservation and Interpretation, Inc. (HCI), to inspect and report on the site. HCI's president, Edward S. Rutsch, and his researchers first studied the city's real estate records. They discovered maps showing the cemetery within the proposed building's footprint. At first, however, they assumed that the weight of earlier structures built on it had crushed and destroyed the coffins, skeletons, and whatever artifacts lay around them.

One small area that had been a street appeared relatively undisturbed. In May 1991, the HCI team dug there in search of the cemetery — and almost at once began to find human bones. They were scattered about in the dirt and rubble, which meant, Rutsch wrote, that they had been dug up before, during earlier construction, and then put back into the earth.

Working in tandem with construction crews, who shored

up their deepening trenches, the archaeological workers soon found undisturbed burials: skeletons, wooden remnants of coffins, even grave-marking stones on what had once been the earth's surface. But no names or other writing.

The more they dug, the more graves they found. The more they found, the slower their work progressed, as they took the time to photograph and document the remains and carefully disinter them. Rutsch needed additional help. An informal group of New York City experts on human remains, called the Metropolitan Forensic Anthropology Team, or MFAT, volunteered to analyze the skeletal discoveries. The MFAT members were experienced in excavating old cemeteries and identifying buried human remains.

At this point, Cantwell and Wall report, the scientists estimated that the cemetery contained about 50 sets of remains.

Old maps that were beginning to come to light suggested, however, that the ABG had been quite large, five or six acres, and so must have contained many more graves. But all the excavation that began in 1991 took place in less than one city block: a patch beneath a street, a gas station, and a parking lot.

The GSA's building contractors, meanwhile, became impatient. They pushed forward with the excavation for the building. Construction, especially for huge structures, is time dependent: Holdups produce down time, which creates cost overruns that, in this case, would cost the taxpayers dearly. Adding to the difficulty, GSA operates mostly outside the limelight. Its officials had little experience in, or willingness to become engaged in time-consuming archeological projects. In the 1980's and 1990's, the feds' major construction needs were courthouses and prisons — workaday buildings for which GSA neither sought nor entertained public fanfare. (But even for GSA, the "Foley Square Project," as it came to be called, was a Big Deal: GSA's largest such endeavor. It cost about a billion dollars.)

Then, in the summer of 1991, the archeologists made a major discovery: More of the cemetery remained than anyone could have guessed. Reason: When 19th-century builders constructed the Victorian structures that were now being demolished, they had leveled the ground. *Not*, however, by scraping and blasting it down to a lower level. But rather, by building up the lower side of the area. This created much more cushioned fill than had been anticipated. The ground covered many relatively intact graves, at the unusual depths of 16 to 24 feet below the surface.

"No one expected burials that deep!" an MFAT archeologists, Leslie E. Eisenberg, PhD, said later. She is a specialist in physical anthropology and urban archaeology and was to become a board-certified forensic anthropologist. Eisenberg had a research position at Lehman College in the Bronx at the time and was an associate of anthropologist Spencer Turkel, PhD, MFAT's co-chairman. (Later, following 9/11, Eisenberg worked on DNA identification of World Trade Center victims for the Office of New York City's Chief Medical Examiner. For the last decade, she has been on the faculty of the University of Wisconsin, at Madison.)

One of Eisenberg's research interests, which led her to take an active role in the ABG project, was bone infections — which can leave indelible marks on skeletal remains — and the relationship of these infections to illness and death in historical and ancient people. Her job on site was exhumation of the remains.

Eisenberg said she started to work on the ABG project late in 1991. It was the backhoe and payloader operators, she said later, who first became alarmed by the unexpected graves they were destroying. They reported this to their bosses, to GSA

SANKOFA?

— and ultimately to the wider world.

On the morning of September 30, 1991 the scientists were working in a 16-foot-deep trench when they uncovered the dark outline of a grave in the red, glacial clay. Then a rusted coffin nail. A coffin lid. And then, underneath that, the next day they found the intact skeleton of a man — in relatively good condition: a black man who had been buried some 200 years earlier. The researchers knew, wrote archeological conservator, Gary S. McGowan and his co-author, Joyce Hansen, in their book *Breaking Ground, Breaking Silence*, that this "was one of the most important archeological discoveries of our time."

In October, GSA announced the rediscovery of the cemetery at a press conference at the site. David N. Dinkins — the city's first, and thus far only, black mayor — declared:

"Two centuries ago, not only could African-Americans not hope to govern in New York City. They could not hope to be buried within its boundaries!"

The diggers uncovered a few stone burial markers, none of which retained any indication of who was interred below. Some graves had been outlined on the surface of the ground with rows of small, water-smoothed stones.

As more bodies were unearthed, estimates for how many had been laid to rest in the Burial Ground rose into the hundreds and then into the thousands. Most — 90 percent — were wrapped in shrouds and buried in wooden coffins. Most of the wood had since completely decayed. The time needed to carefully exhume the remains rose, too. It sometimes took several days to lay bare, document the conditions, and excavate a single grave.

Besides bones and bone fragments, little else remained. Big enough bits of the coffins were rescued for botanists to identify their materials: mostly cedar, pine, and other soft woods. Iron nails survived. So did hundreds of small copper straight pins from shrouds and clothing. *Ditto* buttons; some people were buried in their ordinary clothes. Shrouds, *winding sheets*, and clothing were almost gone, save for tiny patches around buttons and other metallic objects that had kept them from rotting away. There were a few other objects, including clay pipes, jewelry, pocket knives, and coins.

GSA officials became more and more impatient. They estimated that removing 200 sets of remains would set them back four months — and cost an extra $6 million. To speed the process, GSA considered the possibility of ending the slow, careful, scientifically valid disinterment methods and switching to a faster but less scrupulous "coroner's method". Bones would be dug up and dumped on the ground, helter-skelter. Then they would be collected in bags or boxes for removal.

When remains are exhumed in this way, the bones' attachment points are destroyed and the depth and orientation of the body — in what position it was buried and which way it faced — are lost. *Grave goods*, burial artifacts, are scattered. The clues researchers need to identify the bodies and infer their conditions in life are irrevocably lost.

The skeletons' integrity would be destroyed. Demeaned in life, these peoples' remains now would be dishonored.

Leaders of the black community reacted with disbelief and anger. New York State Senator David A. Paterson declared: "It's bad enough that some of the bodies ... were discriminated against in life. But now they are being discriminated against in death." *

To cope with the problem, Paterson formed a task force to oversee the ABG work. It included a wide cross section of concerned New Yorkers — blacks and whites — and preser-

---

* Paterson became the first black governor of New York State in 2008.

vationists, including archeologists Cantwell and Wall, to monitor the excavation and, later, the treatment of the remains. But problems persisted; aggravation increased.

The GSA quickly backed off consideration of the crude, coroner's method of removal. The agency's regional administrator, William Diamond, told the *New York Times* (December 6, 1991, late edition) that "there will be no speeding up that will endanger the artifacts.... [T]he construction will have to take second place. The [construction] dollars will not drive this project."

To understand the events that followed, it is necessary to look briefly at changes that had recently taken place in the treatment and management of American historical sites, including old graveyards.

POPULATION OF MANHATTAN, 1698–1786
(Source: GSA)

CENSUS YEAR	TOTAL	EUROPEAN	AFRICAN
1698	4,937	4,237	700
1703	4,375	3,645	630
1712	5,841	4,886	975
1723	7,248	5,886	1,362
1731	8,622	7,045	1,577
1737	10,664	8,945	1,719
1746	11,717	9,273	2,444
1749	13,249	10,926	2,368
1756	13,046	10,768	2,278
1771	21,863	18,726	3,137
1786	23,614	21,507	2,107

DAVID ZIMMERMAN

2

PROCEDURES

THE AFRICAN BURIAL GROUND (ABG) discoveries and the chain of events that followed did not progress willy-nilly. They were — and are — enmeshed in a regulatory process that defines what may and, more importantly, what may *not* be done with a site and its human remains. This process, the National Historic Preservation Program, promised money with which to proceed.

The program is mandated by federal laws, the best known of which is the National Historic Preservation Act of 1966 (NHPA). The law's purposes are to protect and preserve important American historic objects, buildings, and sites and to provide the public, including specifically Native American groups, with access and influence in the decision-making process on how this is to be done. America's "spirit and direction ... are founded upon and reflected in its historic heritage," which, the law adds, "should be preserved ... to give a sense of orientation to the American people."

The principal impetus for the preservation program was the need to protect significant U.S. historical and architectural sites. A later federal law, the Native American Graves Protection and Repatriation Act of 1990, specifically lists human remains and funerary materials buried with them as "cultural items" worthy of protection.*

At the national level, these NHPA-mandated activities are conducted by the Advisory Council on Historic Preservation (ACHP). It is an independent federal agency, reporting to the President and Congress. It would need to be convinced to intercede on the ABG's behalf.

The ACHP has oversight of historic objects and places that are, or may be, eligible for listing in the National Register of Historic Places. Such a listing qualifies sites and objects for NHPA protection. The ACHP's jurisdiction extends to most objects and sites, including cemeteries and their human remains, that are being financed or developed by federal agencies. A site where a federal office building is being built by the General Services Administration (GSA) is an obvious case in point.

Under NHPA, a federal agency starting a project that could impinge on archeological or historical sites must notify the ACHP and the state historic preservation agency. The federal agency must initiate a review to assess possible collateral damage and to suggest ways such damage might be limited or prevented. Moreover, if possible, the site should be more or less preserved. Local environmental and historical groups are

---

* "I don't care whether this burial is an Abenaki [Native-American], Franco, Irish, or a Jewish one," an American Indian activist declared after an Abenaki cemetery was dug up in Vermont. "We are talking about the desecration of a people's ancestors . . . a disgrace . . . ."

entitled, under the law, to participate in this process; *ad hoc* groups concerned about the government agency's plans must also may be invited in. The gist of these requirements is stated in the NHPA, in a key part called Section 106, and in subsequent federal regulations based on it:

The Act and its regulations stipulate a rigorous process to decide whether an undertaking qualifies for protection. They also prescribe measures to ensure that every interested party has input and that good-faith efforts are made to compromise on any disagreements. The interior of an historic building may be carefully removed, for example, before the building is destroyed. The interior and artifacts associated with it may then be recreated in a museum. The federal agency that damages or destroys the site must pay for this work, at up to one percent of the projected total cost of the new construction. In light of the project's importance and the difficulties that were being encountered, GSA eventually doubled the ABG set-aside to two percent. The federal office building on the ABG site was budgeted at $276 million, indicating that up to $5.5 million might be made available to mitigate the cemetery's destruction. However, no provisions were made for dealing with human remains; ACHP had said it didn't believe any were there.

The parties must commit their understanding to a legally enforceable Memorandum of Agreement (MOA) that stipulates how they will proceed if historic resources might be impacted. A MOA had been signed between GSA and ACHP in 1989. It stipulated that GSA, in collaboration with the National Park Service, was to develop a treatment plan for the site. The New York State Historical Preservation Organization (NYS HPO), which held the state's jurisdiction over the site, had refused to sign the MOA, the ACHP says, because,

SANKOFA?

it claimed, GSA had not negotiated in good faith regarding design issues about the new building. What is more, ACHP says, GSA had not told it anything at all about the planned federal building.

During negotiations, the feds may decide to build elsewhere or they may decide to go forward with their plans. In the latter case, the parties may agree on *mitigation*, i.e., making the best of a bad situation.

THESE STUDIES AND NEGOTIATIONS require input from professional researchers. The tasks are outside the ambit of university scientists — who traditionally *study* but often do not *preserve*. Also, because the time, manpower, and cost tend to exceed university researchers' resources, the need for this work has spawned a whole new class of professional scientists. Their work is called *contract archaeology, salvage archaeology,* or *cultural resource management* (CRM). These scientists are widely employed. The search for missing service men and women lost in Asia and Europe and in all U.S. military conflicts worldwide, is conducted by contract archeologistss working with military recovery teams. They unearth, study, manage, and report on sites like the ABG. The majority of American archeologistss work in cultural resource management.

Some contract researchers work alone or incorporate as small businesses. Others are multimillion-dollar firms. The ABG project engaged one of each type: first, New Jerseyite Edward S. Rutsch and his tiny Historical Conservation and Interpretation (HCI) Company; later, the West Chester, Pennsylvania firm, John Milner Associates.

Most American archeologistss and the lion's share of their research expenditures are paid through federal CRM funds, rather than by universities. CRM work may be less rigorous, and the data may not be analyzed in a substantive way. The ABG scientists planned to do much better.

The old cemetery was dimly known to a few New York City historians. It was marked on 18th-century maps. So, decades later, it was assumed that the cemetery had long since vanished. No provision was made for it in the MOA.

An amended MOA thus was needed to take into account the newly discovered human remains. One was drafted. According to GSA, it was executed late in December (1991). What it did not say is that GSA was empowered to proceed with the project whether or not ACHP and the relevant local agencies approved it, New York City archeologists, Amanda Sutphin, RPA (Register of Professional Archeologistss), explained later.

The GSA, what is more, had "failed to advise" ACHP and other preservationists that a federal office building, serving both old and new federal courthouses, was to be built on the cemetery site, ACHP later reported. ACHP said its officials and its counterparts from New York State had learned of the planned 800,000 square-foot edifice — including 355 underground parking places — in 1991, when GSA issued a draft Environmental Impact Statement. In it, GSA indicated that it had not consulted ACHP about the project, as required by law, because the NYS HPO had agreed that the building would not impinge on historic properties. (However, that office told GSA that it had no record of such an understanding.)

"The misunderstanding," ACHP wrote, "resulted in GSA's failure to address impacts of this project on . . . the African Burial Ground."

Anticipating no problems, GSA's contractors had begun

SANKOFA?

to excavate the site. But archeological testing soon turned up skeletal remains. They were quickly identified as ABG burials.

Rutsch and his crew, along with the New York City academic scientists of the Metropolitan Forensic Anthropology Team (MFAT), began disinterring the remains. By early 1992, they had documented and excavated dozens of graves, whose contents they transported to MFAT's small suite of rooms uptown, at Lehman College in the Bronx, a part of City University of New York, to await further developments.

Discovery of the ABG and its imminent peril created intense and rapidly widening interest, as well as worry among many New Yorkers. The rising controversy drew the media spotlight. This fascination built on two earlier discoveries, one a mile from the ABG, the other, a black New Yorker's dramatic quest to discover his "roots".

The nearby discovery was made where Wall Street runs into the East River in Lower Manhattan. The shoreline there has been incrementally extended out into the water over the last 350 years by the construction of bulkheads — walls — planted in the water; the walled-in areas then were filled with soil, rocks, garbage, and other debris.

This fill, unfortunately, tended to dissolve and slide away into deeper water. To hold it back, New York builders tried to anchor it in place with large, heavy objects. For example, the hulls of derelict ships were moved into position, then filled with rubble and sunk.

In 1981, excavators, with archeologistss in tow, were digging a foundation hole for a large bank building near the East River when they came upon some of these ancient bulkheads and piers. On January 6, 1982, they discovered the wooden structure of a blunt-bowed merchant ship; uncovered, it turned out to be 92 feet long and 25 feet abeam. Its ribs and parts of its hull were clearly visible in the surrounding muck. Tropical shipworms found in its hull indicated that it had been engaged in trade, perhaps in slaves, between the British West Indies and New York.

"The excavation of the ship . . . generated enormous excitement even among the most jaded of audiences, resident New Yorkers," archeologists Anne-Marie Cantwell and Diana diZerega Wall recount. "On a freezing cold Sunday in February, more than eleven thousand of them waited in line to file past the site so they could see the ship."

Further afield, but closer to heart for many black New Yorkers, was the publication in 1976 of the autobiographical account *Roots*, by Alex Haley. He claimed to have traced his family tree back to a village in West Africa and to a young man named Kunte Kinte, who was captured in about 1750 and brought to America as a slave.

This reconnection with Haley's past — his origins — struck a heartfelt cord in many, many Americans, black and white. It was an epiphany to realize that one black man had recovered his family history back to the Old Country. This genealogical quest is a valued achievement for many Americans, humble and bold, including several recent presidents.

Eventually, however, investigative reporters sifted carefully through Haley's story. They showed that it was a hoax. It never happened. Haley hadn't found his personal roots. But by then, "Roots," the story, had attained a life of its own, against which facts could not prevail. It was the most watched US TV series ever, with an estimated 36 million tuning in, in 1977.

SANKOFA?

HISTORIC SITES IN THE VICINITY OF THE AFRICAN BURIAL GROUND

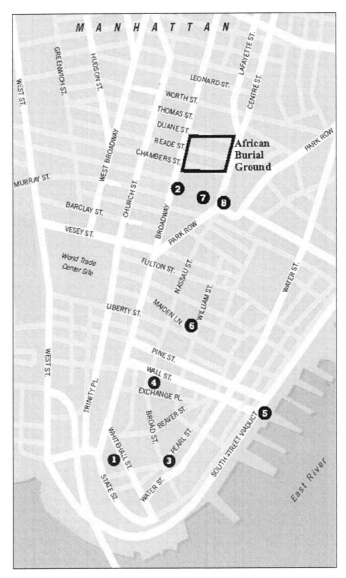

At Civic Center. African American historical sites have been superimposed on contemporary map of Lower Manhattan. They include Fort Amsterdam (1), Broadway (2), Fraunces' Tavern (3), the Wall (4), the Wall Street Slave Market (5), the 1712 Slave Revolt (6), the 1741 Executions (7), and the road to Harlem (8).

David Zimmerman

Subsequently, the human impulse to recover their past grew among black New Yorkers. The ABG beckoned as just such an opportunity. But black people were bedeviled because the federal government, which long endorsed the slavery that brought their ancestors here, now was ready to destroy their bones. This situation anguished — and deeply angered — many of these New Yorkers. Their efforts to preserve and honor the remains reinforced their own sense of being treated as no-accounts by a hostile white society. America's brutally racist history was being re-enacted — and they and their forebears were again in the cross hairs. They were growing more and more furious.* Blakey said:

> [T]he remains.... represent the oppression suffered by their ancestors.... a past that continues to haunt the present. The location of the burial ground outside the wall of the 18th century city reminds African Americans of the lack of respect their enslaved ancestors suffered during the very years their labor was being used to build the city. The unexpected presence of the burial ground beneath the financial and governmental halls of the 20th century city symbolizes for [them] their forgotten role in the creation of the city and the continued disregard for their interests. As construction activities desecrated... the cemetery, the community solidified its resolve to participate in and ultimately control that... exposed ... portion of the original cemetery....

At public meetings, hearings, and sidewalk rallies, black New Yorkers demanded that the excavations stop. Some wanted the digging to cease completely, so that the site could be remade as a national monument to slavery's victims.

---

* Malcolm X had said a quarter century earlier: "Our cultural roots must be restored before life (incentive) can flow into us; because just as a tree without roots is dead, a people without cultural roots are automatically dead."

SANKOFA?

The GSA, as required by the NHPA, convened forums and held town meetings where the black community's views — and others' — were aired. GSA listened. But it did not alter its plans or slow the excavation. The MFAT archeologists and anthropologists struggled, meanwhile, to stay one step ahead of the bulldozers. They were documenting and disinterring 10 to 20 graves a week by working 12-hour days, 7 days each week.

MFAT and Ed Rutsch of HCI had another serious problem: They had no plan. They hadn't decided what to do with the remains they were hastily retrieving.

They were legally required to have a *Research Design*, approved by appropriate governmental agencies, before they touched the first bone; yet they already had removed and stored thousands of them. Rutsch, who was responsible for producing such a plan, was elbow deep in dirt, working at the site with his assistants. So, in the rush to get the bones out of the ground, HCI went forward without a plan, a process that the chair of New York's Landmarks Preservation Commission, Laurie Beckelman, likened to "driving a car in a foreign country without a road map or destination."

There was a second, persistent problem at the dig: While there were a few black people on the scaffolds and in the trenches, almost all the project's leaders — HCI and MFAT — were white. Despite their dedicated labor, they were beginning to be seen by members of the black community as disrespectful foes, enemies rather than allies.

"The fact that in the New York area there are few African-American anthropologists — in fact, none — became our problem!" MFAT chief Spencer Turkel exclaimed.

The fascination and worries that the dig was generating permeated New York City, with many people taking sides: white researchers and government builders *versus* angry blacks and their antecedents' tragic remains. With increasingly critical media coverage, particularly in black and alternative outlets, it seemed a fair guess that trouble was brewing.

3

A DEFINING MOMENT

AS MORE AND MORE bodies were discovered during the winter of 1991-92, relations worsened between the General Services Administration (GSA) and its building contractors, and Ed Rutsch and his scientific excavators. The former were anxious to press forward. Construction contracts stipulate deadlines, and fines are assessed against contractors for each day they're late in completing their tasks. The scientists, however, who now were taking a day or more to excavate each grave brought into view, needed to go slowly — and wanted more time — to remove the remains carefully.

Anthropologist Leslie Eisenberg and her fellow diggers "worked on our bellies" to reach into the graves. The coffins and their human remains were "very wet and very fragile," she recalled. What is more, in some places the coffins were stacked five and six deep, sometimes, confusingly, in different patterns.

Following standard procedures, the scientists documented each site with photographs and painstakingly measured data. Each burial was numbered. The bones were wrapped, individually, in newspaper, and each set of remains, along with the "pedestals" of earth underlying them, was sent by van or truck to the Metropolitan Forensic Anthropology Team (MFAT) headquarters, uptown in the Bronx. MFAT had only three rooms of assigned space there, and they were soon filled with skeletal remains. It was an untidy situation.

Problems continued to arise. Relations between the Descendant Community and the federal contractors grew more contentious. In February, a contractor's backhoe operator was digging a trench to place a footing for the new building when he scooped up and destroyed 20 graves. GSA said it was an inadvertent error, because engineers were using an out-of-date map that showed the spot as free of burials. Community spokespeople argued that the desecration was deliberate at worst, thoughtless at best. They also faulted the archeologists for desecrating the graves of "our ancestors."

They wanted the work to stop. They demanded that the site be declared a national landmark, with a memorial and explicatory exhibits on site. They insisted that the dead, now stuffed in cartons in the Bronx, be reburied where they long had lain.

The two sides were proceeding on a collision course, with the archeologists caught in the crossfire.

Then the game changed.

The first indication came on January 7, 1992. GSA sponsored a public forum on the African Burial Ground (ABG). One participant was anthropologist, Michael L. Blakey, PhD, up from Washington, D.C. He was a research associate at the

SANKOFA?

Smithsonian Institution and an associate professor at Howard University in the Department of Sociology and Anthropology.

According to a chronology that GSA issued later, Blakey presented a proposal entitled "Research Design for Archeological, Historical, and Bio-anthropological Investigations of the African Burial Ground." Blakey sensed a huge opportunity in the ABG and its remains: For racial pride and self-esteem — to illuminate and honor black peoples' mistreatment in the North and their hidden but essential role in building America. For scientific discovery. For historical connection and political advantage for black people.

The ABG population is unique by virtue of its size, age, and location, Blakey wrote, several months later. It "is of unparalleled significance to America's African heritage . . . It is one of the most important archeological sites in this country today in that it is the earliest large skeletal population ever examined through careful scientific excavation. The ancestral remains . . . are also of great spiritual and inspirational significance to the African-American community."

To start the vast research undertaking these opportunities offered, three steps were needed:

- The remains, once recovered, must be moved to Howard University, in Washington.
- Ten million dollars would be needed to do the job.
- Michael Blakey, as a black scientist, must run the project.

A Washington resident, Blakey had tenure at Howard. He held a doctoral degree in anthropology from a prestigious university and was one of only a half-dozen black American anthropologists to have achieved this high level of academic success.

MICHAEL LOUIS BLAKEY WAS born in 1953. He comes from a learned, care-giving family. His father was a dentist; his grandfather, a medical doctor. Michael Blakey also says he is descended from Native Americans — the Algonquins — on his mother's side.

He came early to history and science, when a relative introduced him to what became his favorite childhood hobby: collecting arrowheads, spears, and other Algonquin artifacts in rural Delaware. He also collected fossils and insects.

Blakey prepared exhibits of Native American artifacts and fossils from the shores of Chesapeake Bay, in Maryland, for his school's annual science fairs. One year, he won the grand prize. Off-shore vistas also caught his attention.

Later, his father, Katus Blakey, helped him obtain a summer job at the Smithsonian Institution, working under Donald J. Ornter, PhD, a leader in the field of physical anthropology. (*Bioanthropology* originally was called *physical anthropology* when fossils and bones were the only study material available. With the advent of genetic studies and other analytic methods in the 20th century the definition was widened to *biological anthropology*.)

Blakey examined 50 ancient skulls in the Smithsonian collection. According to *Current Biography Yearbook* (2000), he discovered that the Pueblo people, in the Southwest, developed many dental cavities from eating corn — which is rich in sugars — as a dietary staple, while Native American Suroque people in the East, who he said lived predominantly on oysters, had far healthier teeth.

Blakey acknowledges that he spent more time in high school learning to play the guitar and studying African and African American cultures than he did with science. He went to college nearby at Howard University, intent on becoming a composer. But an anthropology course brought him back to his earlier interest in bygone people. In his junior year, he went on a dig to Belize, in Central America, where he helped map and excavate Mayan villages. Then, he says, he decided to become an anthropologist.

He went to study for his Master's degree at the University of Massachusetts at its main campus in Amherst, Massachsetts (UMass), where he continues to be remembered warmly by a classmate and by an administrator who shared time with him there. Associate provost Esther M. A. Terry, PhD, a specialist in Afro-American literature, and Robert Paynter, PhD, an anthropologist, recently recalled Blakey cordially — and admiringly — over lunch at the university's faculty club. He was quiet and attentive, they agree. He also was exasperated by the white racism he encountered at the university and the wider world beyond. Paynter added later: "He was one of the most even-keeled people I have ever known."

Blakey's feelings and interests, and the impetus to resolve them, brought him into close and candid contact with anthropology professor, George J. Armelagos, PhD, according to Paynter. "Race" was a major topic in the profession then — Was race real? If so, in what sense: culturally? socially? biologically? Paynter added that Armelagos's belief, which Blakey came to share, was that biologic traits can be, and are, influenced by environmental factors, such as dietary restrictions, which can be identified from old or ancient human remains.

This knowledge could then be used to shape change — melioration — in the world at large, because in this view, academic science and real life are linked by a two-way street and so may influence each other. This was a compelling notion for Blakey, who sought an *activist* role in the race politics of that era.

Blakey made a key professional contact at UMass: anthropologist Johnnetta B. Cole, PhD. She is a pioneering black female scientist who became a leader in her field as well as a renowned educator. She served as president of two colleges, Spelman, the historically black women's college in Atlanta, Georgia, and later Bennett College for Women, in Greensboro, North Carolina — where she recently has been succeeded by Esther Terry from UMass Amherst. Cole was appointed director of the Smithsonian Institution's National Museum of African Art. Throughout her long career in academia and public service, Cole has been a strong exponent of civil rights for women, blacks, and members of other minorities.

Professor Cole was born into a well-off black family in Jacksonville, Florida, in 1936. Her father, an insurance man, is said to have been the first black American to earn a million dollars. *Public service* was his mantra. Johnnetta was a precocious learner, entering Fisk College in Nashville, Tennessee, at the age of 15. She went on to study at Northwestern University, in Evanston, Illinois, where she earned her doctoral degree in 1967.

A few years later, she was hired by the UMass anthropology department. She also served, voluntarily, as an associate provost for UMass to recruit qualified black teenagers as students.

Provost Cole and graduate student Michael Blakey hit it off, and she became his mentor and friend. Later, when he had earned his Master's degree, and had set to work on his Doctorate, she co-chaired his thesis committee.

In the thesis, he thanks Cole for her "careful attention to [my] manuscripts." Blakey says: "[She] held the keys to the doors of a new cultural anthropology, a rigorous political economy, and to the ethics of social responsibility which inspired in my studies the traditions of an Afro-American culture of struggle, essential to us both."

They have remained in touch, Cole said recently. When the UMass anthropology department held a party to celebrate Cole's 70th birthday, Blakey participated in the festivities, two other guests recall.

Blakey had become a political activist — a left-leaning black activist — during his student years. This led him to a manipulative sense of anthropology as a career, vis-á-vis its objects of study and the public. In a 1983 presentation to graduate school colleagues — anthropologists and archeologists — he defined archaeology as an "ideological *industry* that produces ideas about ourselves." Archeologists labor to extract and assign meaning to artifacts — "ancient garbage" — from the past. In other words, these artifacts lack intrinsic value; instead, archeologists create their meaning.*

Doctoral degree in hand, Blakey returned to Washington. He taught at Howard and also was the custodian, or *curator*, of its W. Montague Cobb Anthropology Laboratory. He was in charge of a collection of more than 700 skeletal human remains, most of them black people. Dr. Cobb, a physician and an anthropologist, had collected them — mostly from medical school dissection labs — for preservation and study. Blakey was salaried by Howard for this work.

By the time of his visit to New York in 1992, Blakey said, he had published "extensively" on skeletal biology, particularly on teeth, as well as on paleopathology, African American biochemistry, medical anthropology, and the philosophy of science. These reports had appeared in "leading anthropo-

---

* In this, the boundary between archeology and the rest of anthropology is indistinct. Anthropologists study people; archeologists study social and cultural structures, including the ideas, beliefs, and buildings, through which people live. Human remains buried by custom in a cemetery clearly belong to both scientific realms.

logical journals," including the *American Journal of Physical Anthropology* and the *International Journal of Anthropology*. He also had been president of the small Association of Black Anthropologists, a part of the American Anthropological Association.

His research focus — perfect for the ABG project — was historic African American skeletal populations, particularly, methods to determine the ages of fetal and childhood biological stresses on the basis of developmental rings in tooth enamel. Blakey had one other, supreme qualification for leading the ABG research, in his own view and in that of black leaders whom he had met in New York: He was black.

In March, archeologist Edward Rutsch of Historic Conservation and Interpretation (HCI), submitted a Research Design to GSA and the other involved agencies. They took it under consideration.

On March 9, Blakey returned and rejoined the fray. "I inserted myself in the process," he said.

He arrived in the trenches soon after MFAT anthropologist Eisenberg. She recalled later that he gave the impression that being a black anthropologist from Washington gave him special cachet.

She and Blakey soon clashed.

"We [MFAT] were on site when he came," she remembers. "He knew me. He didn't know anyone else or the team . . . . It was clear that the political atmosphere was about to change. It got really down and nasty!"

Blakey claimed, she said, that bone infections on skeletal ankles were caused by shackles, the individuals having been shackled in life. She disagreed, based on her own expertise.

"I don't approve of trying to match observations with beliefs for which there was no evidence at all." Later, Eisenberg said, "You couldn't go to court with something like that!"

SANKOFA?

Eisenberg did agree with Blakey that GSA "never did due diligence" to the ABG and "was not taking its federal historical preservation role [under NHPA] seriously at all."

Blakey went to a meeting of GSA officials and members of the black community. He told them he intended to bring Howard University into the ABG project.

"It was quite a surprise!" MFAT's co-director Spencer Turkel, PhD, said later.

Turkel's hope was that Blakey would work for MFAT, and for the balance of Blakey's visit he introduced him as part of the MFAT team.

"I just said, 'Uh-huh!'" Blakey said later.

The high point of this trip was a visit to MFAT's offices in the Bronx. He was instantly disparaging of the disrespect he saw there.

He was upset, he said, to find that the disinterred bones had been wrapped, individually, in newspapers, including the comic pages and a front page from the tabloid *New York Post*. Worse, the damp bones had absorbed ink from the newsprint so that fragments of headlines could be read, in mirror image, on the bones.

"You wrap *fish* in garbage paper!" Blakey exclaimed.

However, MFAT's Turkel claimed he saw Blakey himself, at the ABG site, using newsprint to wrap exhumed bones. Newspaper long had been the standard wrapping for brittle old bones, Turkel said. Only recently had some conservators claimed that this cheap paper might emit acids that could damage the bones.

"We didn't know that," Turkel acknowledged.

Blakey demanded that the fragile bones be rewrapped in acid-free paper.

Much later, it would turn out that Turkel was right in the

first place: A scientific advisory panel declared: "There is no need to wrap bones in acid-free paper when they will be reburied."

By early May, Blakey was confident enough of triumphing over MFAT that he began writing letters to colleagues soliciting research ideas and seeking associates to work with him on the ABG project.

By early July, he could write to them again to say, in part:

- "[O]ur competitors, MFAT/Lehman [College] have steadily lost credibility with the [black] community because of their dishonesty and ignorance of African-American studies . . . John Milner Associates [JMA] is taking over the archeological excavation."

Blakey had become familiar with JMA when he worked with it on the tail end of an excavation and research project at the First African Baptist Church Cemetery, in Philadelphia. He now told potential ABG co-workers: "Milner is a much better bunch than HCI, and we are now much closer to . . . an organizational relationship with them that will give us oversight of the entire biohistorical analysis phase . . . There is little question that we will be doing the work . . .
- "Our general scheme will probably be approved . . . during the next three months . . . .
- "[T]he African-American community in New York *loves* our [research] design . . . [emphasis in the original].
- "[S]ome of the archeologists [there] continue to resist efforts to have the remains . . . removed from the city. They will not prevail . . .
- "Despite all of the deviousness on the New York end, rest assured that I will keep everything clean on ours."

SANKOFA?

In midsummer, matters came to a head between New York City and the feds: GSA told the mayor, David Dinkins, that the agency intended to remove 200 more burials. Dinkins wrote to William Diamond, GSA's regional director, formally requesting that excavation stop until plans were firmly in place for the remains' disposition. GSA's "continuing actions," he said, "are deeply disturbing to many New Yorkers."

Diamond refused to stop. Since GSA was in compliance with federal regulations, he claimed — albeit it wasn't — the work would go on.

Anger erupted.

Congressman Charles Rangel (Dem., New York) spoke — eloquently — for the city's black community:

"This is our Ellis Island. This will not be business as usual!"

There was one source that could stop the excavation: the US Congress, which had ordered the building and had appropriated the money to erect it. Providentially, Congress now stepped in. The chairman of the subcommittee on public buildings and grounds of the Committee on Public Works was Augustus Savage (Dem., Illinois). Gus Savage is black. And he was furious. He called a hearing for his subcommittee in Manhattan, across the street from the ABG. It was held on July 27, 1992. All parties testified. Savage's ire continued to rise. At the hearing's end, he shouted at Diamond and his GSA associates:

"Don't waste your time asking this subcommittee for anything else as long as I'm chairman, unless you can figure out a way to go around me! *I'm* not going to be part of your disrespect."

That stopped GSA in its tracks.

The agency quickly announced it would exhume only remains that already were exposed. Construction for the tower building, where the burials already had been removed, would go forward. But a planned pavilion, where 200 other sets of remains were buried, would remain unbuilt. The graves would not be disturbed.

Excavation of the ABG had ended.

4

GOING FORWARD

THE CONTRACTORS' BACKHOES AND bulldozers fell silent. Congressman Gus Savage's shot across the General Services Administration's (GSA) bow now forced its officials to agree to go forward in a procedurally correct fashion. This meant that the African Burial Ground (ABG) and its denizens' bones had become a project in their own right — soon, a *federal* project — with still-to-be-set trajectories, under GSA's overall control.

Excavation would go forward within 290 Broadway's original footprint. But the agreement Savage and GSA worked out canceled the agency's plan to build a four-story pavilion atop the remaining open area. It was to have housed an auditorium and a day-care center. Instead, this corner of the cemetery would be preserved as a memorial site for the people buried there — and as a place for the exhumed remains' eventual reburial.

Even in that fragment of the burial ground, however, excavation for the placement of the building's huge concrete footings continued to destroy graves. A dozen sets of remains that had previously been exposed were later disinterred and added to those already held at Lehman College in the Bronx. But no further exhumations were permitted.

The excavated area is 9,500 square feet. Archeologists estimate that this represents only 3.6 percent of the cemetery's original 6.6 acres, which ran downhill from what is now Broadway, along present day Duane, Reade, and Chambers Streets to Foley Square. The 400 disinterred remains, they add, are likely to be less than three percent of the total number of burials.

What was to be done with these remains? Under the agreement worked out between Rep. Savage and GSA's top official, Richard G. Austin, this and other decisions were to be thrashed out by the Advisory Committee on the ABG project that Mayor David A. Dinkins had set up a few months earlier. The committee would then forward its decision to GSA.

The Mayor's Committee would be reconstituted by Congress as the Federal Steering Committee (FSC) for the ABG. It was led by historian Howard Dodson, MA, chief of the Schomburg Center for Research in Black Culture, a part of the New York Public Library, in Harlem. Dodson was an expert on the comparative history of African peoples' enslavement in the Americas. He had played, and would continue to play, a leadership role for the Black Community in representing its concerns about the ABG project's direction.

THE MAYOR'S COMMITTEE WROTE a succinct plan for the project, which was presented to GSA and Congress in September. The goals and objectives included:

SANKOFA?

> *"To correct past injustices and indignities by creating appropriate memorials including a monument, or world-class museum, art works and interpretive displays commemorating the 18th Century Africans who were buried in [what was then called] the 'Negroes Burial Ground'. . . .*
>
> *"To preserve and reinter the exhumed human remains in the burial ground site in a dignified and culturally appropriate ceremony."*

A third goal was more controversial:

> *"To ensure the development of a culturally sensitive, scientifically credible research process under the direction of African-American research scientists."*

This designation ruled out the Metropolitan Forensic Anthropology Team (MFAT), whose eight members included no blacks. Also, this "racial" precondition for scientific appointment was — is — highly unusual; "race" is not a scientific credential. What is more, such a preference or precondition for a federally funded job was patently illegal under federal law. It is ironic that hard-won civil rights initiatives to open academic and institutional positions to blacks and other minority group members were now being used by blacks to deny equal job access to others.

GSA, in fact, had not waited. It had already picked the ABG project director. Not surprisingly, it was Michael Blakey. As Dodson told Congress on September 24, 1992:

> "[He] has been contracted by GSA as the Scientific Director. . . . Blakey has assumed oversight responsibility for all scientific aspects. . . He will play a central role in the development of the long-awaited Research Design."

But Dodson also appears to have been hedging his bets: At a mid-August meeting of the Mayor's Committee, he said, according to the minutes, that "he has received direct and indirect pressure to have Blakey remain on the project. "But Dr. Blakey's continuance will be determined by the [soon-to-be appointed] Federal Steering Committee that would absorb the Mayor's committee."

Going forward on the project, of course, would cost money — and thus far, none had been appropriated. The late New York Senator Alphonse D'Amato, Jr. (Rep.), now introduced a bill to provide $3 million in start up funds.

President George H. W. Bush signed it into law.

TWO IMPERATIVES NOW DROVE the Black Community and its leaders: They fervently wished to honor, and win, wide recognition for their ancestors, with whom many had become profoundly attached in spirit. This meant, foremost, proper respect for the remains; specifically, it meant their return to the earth — their reburial — in a respectful and celebratory way. The ABG, the plot from whence they'd been disinterred, was the preferred reburial site.

The second imperative: Study the remains for what they might reveal about how their forebears lived and died. And also, more generally and from a scientific point of view, study them for what might be learned from such a happenstantial but monumental resource: 400 sets of human remains that were 200 to 300 years old.

The two goals might not be mutually exclusive. But there was strong sentiment in the Community against harming, or even messing with, the remains. Particularly, there was strong resentment at the prospect that these hallowed bones would be handled and harmed by white scientists.

SANKOFA?

"For MFAT to have possession of and control the study of these remains resembles, to the African-American community, what it would be for the Jews to have the Nazis study victims of the Holocaust, of *their* holocaust," Blakey was quoted as saying. The comparison, he insists, is not "literal", but "is effective in communicating the emotional feelings [of] the Descendant Community."*

So, according to this line of reasoning, MFAT's white scientists — few if any of whose ancestors lived or kept slaves in 17th- and 18th-century Manhattan — were guilty of genocide against blacks. This condemnation of all whites, based on real or inferred behavior of some of them, is, of course, racism. *Black* racism, as it were.

There is no doubt that the burial ground's remains stirred strong feelings in black people. Historian Dodson said, for example, that when he first saw the skeletal remains at the ABG, his "scientist-historian" side marveled at what might be learned from them. "But the 'roots' side of me said, 'You're not supposed to mess with the dead!'"

More poignantly, the late black jazz musician Noel Pointer recalled, at a remembrance vigil on the ABG site:

"You hear all about these archeological and scientific terms, and you say to yourself, 'This is very interesting!' But until you get down there, and you see what the bodies look like — you see the skeletons of the children lying on the earth — then you realize that these are people."

A less nuanced objection came from city councilwoman Mary Pinkett of Brooklyn. With tears in her eyes, she told New York's Landmarks Preservation Commission that the

---

* The Descendant Community is not limited to descendants of NY City slaves. According to the *Chronicle of Higher Education*, it consisted "for all practical purposes, of any black American who cared to voice an opinion...."

Descendant Community's work was a declaration: "This is enough!" she said. "You can't walk over the bodies of our ancestor anymore . . . . .

"Our people were sold — into slavery — their children were nothing but chattel. And after you buried them, you didn't give a damn! Not only didn't you give a damn, but you built parking lots over them!"

Alton Maddox, Jr., a lawyer, said:

"So many of our people, because they've been so mistreated in life, really want to rest in peace!"

Blakey, as a black scientist, argued, however, that "to learn from archeological sites constitutes another form of respect for our ancestors. . . . It has special importance because so much of African-American history has been lost."

THE DISPUTE OVER WHO would do this work continued. MFAT did not want to let go of the bones or the stellar research opportunity they provided. Or, perhaps, the money: The figure being bandied about for conducting the research was $9 million.

Financing and all the other options for research and memorialization of the cemetery's population depended on the ABG being designated a landmark under the National Historic Preservation Act (NHPA). The city's Landmarks Preservation Commission took this step several months later.

Civic pride also was an issue. New Yorkers, blacks and whites, were understandably reluctant to see a prize discovery surrendered to Washington, even if only briefly (since the bones would be returned to Manhattan for reburial wherever the research was conducted).

Not surprisingly, given the rising stakes, the conflict between Blakey, who now had the upper hand, and MFAT, which was not ready to give up, grew sharper. And uglier. "Over time, what might have been merely an academic squabble, a dispute over scientific and historical methodologies, has deepened into a bitter conflict," one observer reported. "[O]ne of the issues is race." This was an issue that Blakey knew how to exploit. MFAT's co-director, Spencer Turkel, and his colleagues did not. For them, science was nonpolitical. But there was a strong impression in the Black Community, nurtured by Blakey, that MFAT was holding the bones hostage in its bid for the research contract.

News of the enmity soon became public. The *Village Voice* published (May 4, 1993) a five-page investigation by reporter Karen Cook that was headlined "Bones of Contention." The bones "have been at the center of a prolonged struggle between two scientific camps": Blakey's and MFAT's, Cook wrote. "One of the major issues is race."

Soon after Congressman Savage's successful confrontation with GSA, Blakey paid another visit to the remains' repository at Lehman College. He was accompanied by activist supporters, with video cameras rolling.

It was a hot summer day. The rooms were not air-conditioned. The boxes and file cabinets filled with earth and bones exuded a musty smell.

Blakey sniffed the air.

He caught a whiff of a different odor — one that he recognized: a chemical. It was a solvent used by physical anthropologists and their technicians to remove the mold from moldering bones.

Blakey became incensed:

"They seemed to expect me to look at whatever they were planning to show me. It was almost as though they were expecting a tour, rather than an inspection!"

Turkel saw the incident as a setup. MFAT never tried to hide the fact that some of the bones were moldy, he said, nor that MFAT workers had used solvents on them; this was a standard solution for a common problem.

"He had this whole act," Turkel later told a reporter. "It was a routine: *fee, fi, fo fum,* I smell mold!" And the [Descendant] Community was very impressed that he was able to suss that out."

Blakey and his allies condemned the outrageous way the remains were being treated.* They charged MFAT with racism. "Black people are used to putting together the bits and pieces of implicit prejudice," Blakey declared.

The newly formed FSC seconded Blakey's critique of MFAT incompetence and mismanagement and gave him its nod. The committee recommended that Blakey be appointed scientific director for the entire ABG project. GSA did so on October 1, 1992. What remained undecided was where specific parts of the research would be conducted, and by whom. These decisions hinged on a specific plan for the project, a *Research Design*.

The Research Design requirements, as defined by the NHPA, are: "a statement of proposed identification, documentation, investigation, or other treatment of a historic property that identifies the project's goals, methods, and techniques, [the] expected results, and the relationship of the expected results to other proposed activities or treatments" — which in this case meant plans to honor and commemorate the long-buried human remains.

THE ADVISORY COUNCIL ON Historic Preservation, meanwhile, had examined Historic Conservation and Interpretation

---

* But, Blakey acknowledged soon thereafter, in the publication *Archeology* (1993): "[T]he bones were in good shape."

SANKOFA?

Inc.'s (HCI) Research Design, and found it wanting. The Council asked for revisions. But by then, in early summer, Blakey and Howard University, in conjunction with John Milner Associates (JMA), had submitted the first draft of their design document — and it was clear that they had greater resources and, in Blakey, the correct skin color to advance the project. HCI was dropped out of the running.

HCI's fevered efforts, however, had distressed some interested parties, particularly black New Yorkers. They charged that the lack of planning was just one more example of racial discrimination in the research world.

ON JULY 1, 1992, GSA had appointed JMA to administer the ABG project, replacing HCI, under contracts worth millions of dollars over the next 15 years. JMA had the special advantage of having managed the excavation of two black burial grounds in Philadelphia. What is more, the company was in touch with Blakey: He had worked on JMA's project to study black human remains retrieved from the First African Baptist Church cemeteries there. This project was completed by Blakey's UMass schoolmate, anthropologist Lesley Rankin-Hill, PhD.

Faults had been found in the first version of Blakey's draft Research Design, and over the summer he revised it. He and JMA resubmitted it to GSA and the FSC in mid-October 1992. The ABG's skeletal remains and artifacts, they wrote, are "the only concrete evidence so far recovered of the enslaved Africans who inhabited New York City from its earliest colonial beginnings...."

This draft explores the scientific (and historic) value of the cemetery, as well as its spiritual value as "the Africans' only autonomous social space, the only place where they were allowed to congregate with regularity, in large numbers, and beyond the purview of the authorities."

In a passage that is assertive but also highly speculative, Blakey characterizes funeral gatherings at the ABG as the origin of American black culture [the emphases below have been added]:

"Unmolested by outside interference, the Africans venerated their deceased family and friends at the ABG from roughly 1712 to 1790. Enslaved Africans *perhaps* led other slaves in traditional rites intended to memorialize the dead. Through the performances of these rituals, African ways were *perhaps* transmitted from one generation to the next. For more than two generations the ABG provided a precious space in which a unique African-American society and culture took shape."

Blakey promises that scientific analysis of the remains will provide a broad range of baseline data "against which subsequent changes in the biology of African and African-descended populations can be compared and evaluated."

Chemical analysis of teeth, bones, and hair, he continues, will yield baseline data on toxic elements found in humans who died in America before the rise of industrialization and its inherent home- and work-related burdens of pollution. In this, Blakey proposes a scientific use for the remains that falls outside issues of "race" and ethnography.

The document promises research "on the epidemiological and agricultural/nutritional practices of ... West African ethnic groups." However, it provides scant few particulars on how this was to be done.

Most earlier studies on such remains, Blakey notes, have been "descriptive"; little attention was paid to the "biocultural" and "historical" specificities of these persons' lives. This study, he promises, will "overcome the previous inability to provide "a truer historical picture of Afro-New York...." Blakey's aim was to discover indigenous African traits, whether physical or behavioral, that had survived capture, transatlantic transfer, and acculturation with white slave masters. He saw those traits as conduits to spiritual rebirth and renewal.

Blakey assumes that biological study of the remains will reveal significant socioeconomic and cultural data, as well as evidence of enslavement. Bone specimens taken from the remains will also reveal genetic variations between the blacks buried in the cemetery and other populations, as well as the effects of changes in the New York environment between about 1710, when the cemetery was started, and more recent times.

This was a very tall order.

Remarkably, he does not say precisely how he plans to do all of this, nor does he say precisely what he hopes to find. In this draft Research Design, there are dozens, if not hundreds, of general statements of intent like this, with little or no indication of how Blakey hopes to get from A to B.

Here are two of these universal, find-out-everything "specific aims" for the project:

- "An assessment of African and American cultural origins of individuals, using biological and archeological methods."
- [An] assessment of differences in the physical quality of life in West Africa, southern plantations, and New York using biological indicators of childhood and adolescent health experiences."

In short, Blakey proposes encyclopedic studies, without specifying what they would be.

The science and the spirituality of the project are inseparable in Blakey's view. The remains are "of great spiritual and inspirational significance to the African-American community," he states. So his Research Design "recognizes the necessity of ongoing consultation with religious leaders who will work with the scientists . . . to see to the sacred aspects" of the work:

"Periodic religious ceremonies are anticipated throughout the project."

This conflation of the rational, scientific element of the work with the irrational, spiritual aspect is highly unusual in biologic research. More unusual: Blakey saw himself as leading both elements of this initiative. When science and spirituality are both involved in an inquiry or activity, the two realms are usually represented by separate and distinct individuals or groups. Each has its own — and almost always *different* — assumptions and intentions. Under the best of circumstances, they work in parallel, without getting in each other's way.

At a patient's bedside, for example, the surgeon and the shaman each has a role. One operates. One prays. They keep these patient services separate. If the patient dies, the pathologist or medical examiner has one key role to fulfill: determine the cause of death. The priest and the undertaker, quite another: comfort the bereaved and bury the body. For the ABG, Blakey, with some help from associates, planned to fill both roles.

5

TAKING OVER

IT IS NOT CLEAR exactly when the Metropolitan Forensic Anthropology Team (MFAT) realized that anthropologist Michael L. Blakey had come to New York to supplant rather than support its study of the African Burial Ground (ABG) human remains. But by October 1992, his purpose was patently clear.

October also provided MFAT with an opening to counterattack: When Blakey and John Milner Associates, the proprietary archeological firm with which he planned to collaborate, submitted their 146-page draft Research Design dated October 15, this opened a 60-day period during which the General Services Administration (GSA) would accept and consider written comments from all interested parties. The draft also was sent to members of the Mayor's Committee, newly chartered by Congress as the Federal Steering Committee for the ABG [henceforth, referred to as the "Steering Committee"].

The October draft would soon become better known for what it did not say and what it got wrong — as will be described below. Blakey did not say in it how he planned to determine scientifically from the bones which were from black people and which from whites — and he indicated that he did not plan to try.

The researchers were dismayed when they read Blakey's draft: Metropolitan Forensic Anthropology Team (MFAT) director, Spencer T. Turkel, PhD, and his associates drew up a list of more than 100 colleagues — anthropologists, archeologists, pathologists, and others — and asked GSA to send them copies of the document and solicit their views.

Many on MFAT's list responded. Many sent devastating comments on the Blakey document. GSA subsequently preserved the comments in a bound volume of several hundred unnumbered pages.

Three types of criticism predominated:

- Objections to the destruction of the ABG and to most or all efforts to mitigate the loss.
- Critiques of the planned studies and planned disposition — through reburial — of the remains.
- Objections to Michael Blakey and his plan to ship the remains and shift the research to Howard University in Washington.

The *Don't mess with it!* approach was succinctly stated by a Mr. Carlton Reid, not otherwise identified, in a brief letter to GSA:

> I see the cemetery as sacred ground of our Ancestors, and I strongly feel that it should be left alone by intruders, and [should

be] moved into a huge mausoleum.... [Then] all African people should make daily pilgrimages to this sacred place. I also think each newborn should be taken there for his or her naming ceremony!

Since the present report is an analysis of the ABG project's *scientific endeavors*, reverential objections for the most part must be left for others to recapitulate. But it is worth noting that the proposed study was sharply questioned by people of faith.

Many of these objections are carefully presented "in the name of Alleh, the Beneficient, The Merciful," in a letter written for the Admiral Family Circle Islamic Community in New York by Sheikh [Sheik] Abd'Alleh Letif Ali, of Manhattan. He found the plan "very interesting and informative." But, he said, African American and Islamic communities are not primarily interested in the proposed physical/forensic anthropological studies. Rather they seek information on whether and how the ABG population expressed their African and Islamic heritage. The Blakey draft fails to say how and by whom such studies would be done.

The Sheikh notes, too, that the draft Research Design contains no detailed budget. This is a prescient comment, given the monetary problems that would arise later.

"How realistic in terms of existing funding are the design activities," Ali asks. "Is all the money in place . . . ? From what pot of money will the funds come?"

He notes that the document says that the remains are "primarily" those of Africans. He urges "caution" and "care" to be sure that possible remains of mulattoes and people of mixed black and Native American ancestry not be confused with persons of European origins. Finally, Sheikh Ali asks these thought-provoking questions:

"What are the benefits to the African community in America of the scientific analysis of the [ABG] remains? What is the benefit to science?"

A less sophisticated objection was raised, initially at a public meeting, by Eloise W. Dicks, a member of the Descendant Community. Speaking as an "ancester [sic] to the Afriken [sic] Burial Ground," Dicks challenges "certain paragraphs and lines [of the draft Research Design] that I felt was [sic] not correct, [and were] untrue, offensive and undesirable to the Afriken people.

"I also feel," she added, "that the draft is coming from a European prospective."

One plan had Ms. Dicks' unqualified support:

"Yes, Howard University should be doing all the research!"

A major bone of contention between many of the professionals, whose views MFAT had solicited, and Blakey had to do with the "race" of the ABG remains. In other words, exactly *whose* bodies were buried in the ABG? Who were they?

As the cemetery's name indicated, many — perhaps most — were black people with African origins. Many were slaves — but which ones? And which ones were free men and women? (A related problem: What *kind* of slavery was practiced in New York City? The *chattel slavery* of Harriet Beecher Stowe's Simon Legree in her *Uncle Tom's Cabin*? Or a different, conceivably less-onerous *indentured servitude*, a category that also, not incidentally, included a great many white immigrants to colonial British New York?)

Besides blacks — enslaved or free — 18th-century anecdotal reports indicated that other people also were interred there: white paupers, for example, and American soldiers and sea-

SANKOFA?

men who were captured by British forces and subsequently executed or died in custody during the Revolutionary War. The city's gallows stood near the cemetery. So it also seemed reasonable that some people who had been hanged there — white as well as black — had been buried there, too. And what about Indians — Native Americans — who died or were killed on Manhattan Island? If they, like blacks, were not welcome in the consecrated ground of the Trinity Church Cemetery, a half mile downtown, might not they, too, have been buried in the unconsecrated land of the ABG? *

How, then, decide who was what? Customarily, in Europe and America, whites were differentiated from blacks — on the basis of skin color, or "race." Skins, of course, come in all shades and hues, and "interracial" couplings do dilute "racial purity," whatever that may be. But little, if any, skin remained on the remains — so individuals' skin color could not be determined. Moreover, the concept of "race" was itself problematic.

In the scientific classification of living organisms, or *taxonomy*, each great group, or *phylum*, of beings — including "plants" and "animals" — is divided and re-divided based on criteria of common descent. The penultimate division is into "species". If two look-alike animals live close to each other and naturally mate and reproduce themselves together, then they are members of the same species. If they don't, they aren't.

In the 19th century, scientists and other white people decided that black people — whom many viewed as "savages" — belonged to a different and inferior species of mankind. This differentiation then was used to justify black slavery and seg-

---

* Project archeologists reported later that there is no positive evidence that any of the exhumed remains were Native Americans.

regation: "They aren't people, so there is no moral reason to treat them as equals."

But sex of course did cross "racial" lines: Thomas Jefferson and the enslaved Sally Hemmings were an 18th-century example. The late US Sen. Strom Thurmond (Rep., South Carolina) and his family's maid, Carrie Butler, are a 20th-century example. (Thurmond was 22 when he impregnated 15-year-old Butler.)

Racial or "racist" theories also were used to justify the Holocaust and similar depravity against Asians and others. For these reasons, world leaders acted, over a half-century ago, to quell the concept of multiple human species. In 1950, the United Nations Educational, Scientific and Cultural Organization (UNESCO) declared that World War II had been facilitated by "the doctrine of inequality of men and races." In a Statement on Race, UNESCO said:

"Scientists have reached general agreement that mankind is one: that all men belong to the same species, *Homo sapiens*."

But this did not end the matter for scientists, for racists, or for racist scientists. UNESCO did acknowledge a division of mankind (*H. sapiens*) into three biological "races": Mongoloid, Negroid, and Caucasoid.

Mankind might be one, but there are unquestionably various types of men and women within a species. Scientists classify these difference as *subspecies* and also as *races*.

"This makes a lot of people queasy, for obvious reasons," said the late mathematician and science historian Norman Levitt, PhD, of Rutgers University in New Jersey. He explained that racial differentiation is useful as a "rough sorting device" for separating species into subgroups that have gone through some degree of genetic isolation but have not separated into different species since they still can interbreed. Two races of a species usually inhabit different territories.

SANKOFA?

"Human racial type obviously is a 'fuzzy' category, with indistinct and porous boundaries. It tends to be unstable over time as populations meet and merge reproductively," Levitt said. "But if you don't drive it into the ground, and try to talk about 'pure' racial types, *race* is a viable and analytically useful notion founded on real biological difference... backed up by discernible difference at the genetic level...."

Black Africans enslaved in New York clearly had some biologic differences as well as socioeconomic differences from white colonists, who came — or whose forebears came — from Europe. How could these groups be compared *objectively*, or at least, as objectively as possible? Answer: Based on genetic profiles (to be described in Chapter 18), it was beginning to be possible in the early 1990s to say, for example, that a person's — a set of remains' — ethnic origins were people who lived in Africa, and in one part of Africa, not another. What was more, genetic differences between people — between those of African descent and those of European descent — could in some instances be discovered by differences in genetically determined ratios between the dimensions of certain bones, as well as by differences in the skulls and jaws and teeth.

None of these methods was universally accepted by scientists who study skeletal remains. But one such method had been used by MFAT's Spencer Turkel, and his co-director, anthropologist James V. Taylor, PhD, of Lehman College.

This is a statistical method with an unfortunately long name that can be abbreviated MDFA (multivariate discriminant function analysis). One or more bones can be used. By taking measurements of the hip bone and the thigh bone, for example, Taylor had found he could use MDFA to correctly identify some 95 percent of black men and 90 percent of white women when the bones had been coded and their origins —

black or white — were known.

Taylor and Turkel, what is more, were using MDFA to evaluate the burial ground remains stored at Lehman.

This infuriated Blakey. The method was racist, he charged. It shouldn't be used. But that left him with a major problem: How else was he going to differentiate the remains — people — of African origins from the "Euro-American" population with whom he hoped to compare them? How could he apply biological methods — genetic, chemical, and other studies of the bones — to establish the "baseline biology of the African-American population in the U.S.," which is, he said, "so important for understanding the causes of subsequent biological and health changes!"

Blakey never says, specifically, how he will do this — how he will differentiate black from white. He ducks this key issue, saying that since the bones come from an "African" cemetery, it can be assumed that they're blacks' remains, unless they look like they're not. He planned simply to eyeball each set of bones. Europeans, he said, could be identified by the shape of their skulls. — their "cranial morphology". On this basis, "approximately eight percent appear to be of European descent."

Later, however, Blakey reported that only 27 skulls out of the 400 sets of remains were "sufficiently complete for this analysis . . . [A]ll have facial and head dimensions that are more similar to the tropical peoples" of West and Central Africa "than they are to any other .... populations."

MFAT supporters seized on Blakey's vagaries in their letters of comment to GSA. Forensic anthropologist Alexander Sonek, PhD, of San Diego University, wrote:

> "Modern forensic science . . . deems it necessary to ascertain the

characteristics of race, age, sex . . . of the decedent (specimen) together, and independently of one another, exclusive of the [burial] context in which the remains are found."

Sonek acknowledges that "race" is the most difficult factor to identify. But, since it influences the analysis of age, sex, and other variables, an estimation of "race, if at all possible, must be made prior to other analyses."

In the past, Sonek notes, researchers have too often looked at and measured the skull but neglected to analyze the skeleton — which is also what Blakey planned to do. Now, however, Sonek said, Taylor and Turkel's MDFA has been shown to be valuable in estimating race, age, and sex, particularly when, as is the case with most sets of ABG remains, the skull is shattered or missing. Standard scientific methods, he concludes, require that both the skull and the skeleton be studied to estimate ethnicity, first, and then sex and age.

A physical anthropologist at the US Army's research and development center in Natick, Massachusetts, Madeleine J. Hinkes, PhD, agreed:

"I am greatly perturbed by the apparent assumption that all [the] remains are Negroid. Such assumptions are the bane of good science." She added:

"My first impression is that the project was designed with more of an eye towards 'political correctness' than to good scientific procedure. Why else would the intensive skeletal analysis take place, in relative isolation, at Howard, never known to be a powerhouse in physical anthropology?"

One of Hinkes' colleagues, pathologist J. Michael Hoffman, MD, PhD — a specialist in the analysis of human remains — highlights the contradictions in Blakey's proposal and their consequences. This commentator, from Colorado

College in Colorado Springs, cites these passages from the Research Design:

- "... over 400 remains, *primarily* of Africans ... [Hoffman's emphasis added]."
- "... mixed enclave ... of several European nationalities who lived side by side with Africans/African-Americans (some of whom may be buried in the ABG) ...."
- "... approximately 7% appear to be of European descent."

No data support the 7% estimate, actually 8% according to Blakey. But, accepting that figure means that 32 individuals (8% of 400) in the sample are of European descent.

"Glaringly absent" in the Research Design "is any specific mention of methods used to assess ancestry [race], other than a vague reference" to "the form of the face and other cranial and dental dimensions .... [T]here is no mention of how ancestry will be determined in those cases where insufficient facial and dental remains are present."

Blakey vehemently rejected these critiques, which he disdainfully dubbed "racing." More specifically, Euro-American white racing.

A critical view also was entered by a pioneer excavator and researcher of black burial grounds, Jerome (Jerry) C. Rose, PhD, who was professor and chairman of anthropology at the University of Arkansas, at Fayetteville. Rose, who is white, found fault with Blakey's biases.

Blakey had written, for example, that the ABG was treated with disregard and disrespect that would not have happened with a white cemetery. This is "patently untrue," Rose wrote, since cemeteries belonging to many ethnic groups had been

SANKOFA?

"despoiled" in the name of progress. More importantly, Rose doubted that analyzing teeth could determine an individual's continent of birth, since all slave and lower class immigrants to the Americas have been found to continue their traditional dietary practices while working for their New World masters. Moreover: "What is the evidence for maize (corn) consumption during this period in New York?"

This objection had particular weight coming from Rose, for he, like Blakey, had studied at the University of Massachusetts, Amherst, where a main focus of anthropological study was examination and analysis of teeth and other dental remains as indicators of childhood stress and dietary conditions.

Rose had attended UMass before Blakey, but he said later, in an interview in Fayetteville, that this affinity of professional outlook and interests was common to graduates of anthropology there. Since teeth are the hardest part of the human body, he noted, they may be the longest lasting — oldest — remnants still available for later study. Yet, Rose concluded with regard to Blakey's overall proposal:

"[T]he entire bioanthropological design is sufficiently devoid of specifically designated methodologies to prevent [the] determination of whether the [research] goals are achievable or not.... Thus they cannot be judged for adequacy."

Blakey rebutted such criticism on racist grounds. White scientists, he charged, have historically studied black bones in order to assert whites' power and superiority over black people. This "racing," he and an associate Cheryl La Roche, said in a paper published several years later in the journal *Historical Archaeology* (1997), has been associated with arguments in support of black inferiority, social and biological distance, and stereotypical images . . . ." The MFAT members, Blakey and LaRoche add, "seemed keen on demonstrating to the public their technical knowledge by showing the cranial and [bodily] traits they used to classify the faces of skeletons. Members of the New York Descendant Community often identified these explanations of facial and pelvic traits as troubling.... Why should [one of them] have asked, perplexed, 'How could it be possible for a femur to represent her ancestry?'"

Not all comments were negative. "I find the research proposal to be of the highest caliber, and I have no further suggestions," wrote physical anthropologist Jerry Melbye, PhD of the University of Toronto at Mississauga, Ontario, in Canada.

But among the scientists who commented on Blakey's proposal, 11 liked it and 49 did not — a one to five ratio — according to my own rough survey of the letters. The main objection: Studying the skull, but not the skeleton ignored the best — and only — way to decide whether a set of remains came from a black person, a white one, or an indigenous Native American.

The Steering Committee and GSA nevertheless sided with Blakey. He would do the research. And of course he would do it at Howard. As committee member Miriam Wright declared, "If it was an African find, we wanted to make sure it was interpreted from an African point of view."

Blakey would deny, much later, ever having argued that only black scientists were qualified to study black human remains. But that was the gist of his argument at the time, in 1992-93. "Seizing intellectual control," he wrote shortly afterward, means "that the criteria for competency have been expanded to include an affinity for African-American culture, past and *present*, and comfort with and knowledge of the politics of African descendant communities and their cultures and their histories" [emphasis in the original].

It is doubtful that any one *except* a black person could meet these criteria.

6

COUNTERATTACK

BAD ENOUGH THAT MANY scientists, alerted by the Metropolitan Forensic Anthropology Team (MFAT), disliked Michael Blakey's draft Research Design. Worse, from his perspective, the gate-keeping agencies didn't like it either!

Unfortunately the chairman of the New York City Landmarks Preservation Commission, Laurie Beckelman, wrote to the General Services Administration (GSA) to say, the draft "does not meet the requirements of the Memorandum of Agreement (MOA), and is not acceptable in its present form." It lacked specifics on the time, personnel, and money that Blakey said he needed. The umbrella national "landmarks" agency, the Advisory Council on Historic Preservation (ACHP), agreed: The draft Design "lacks adequate specificity in critical areas which would enable us and other[s] . . . to evaluate the proposed work, and would enable GSA . . . to make decisions regarding appropriate actions . . . ."

"Unfortunately," ACHP archeologists added, our "careful analysis reveals that the document lacks a well-articulated strategy concerning research priorities and proposal treatment and mitigation problems" for the African Burial Ground (ABG) site. The "single most problematic part of the Research Design is its presently incomplete . . . data recovery plan.

"We can see no obvious, integrated plan for the proposed analysis . . . ."

The ACHP staff archeologists could not find a series of pertinent research questions that Blakey and his team hoped to answer. Lacking, too, were a description of the kind of information needed to pose such questions and a description of how the excavated material would be analyzed to answer them.

The archeologists stressed that ACHP does not approve of subjecting study materials "to all kinds of analyses simply because these techniques exist." They concluded ACHP's blistering critique by saying: "In our view, the present Research Design is unresponsive to . . . the public interest . . . . It is too general to be of much use in making decisions or as a guide to direct work on the extremely important ABG site."

The panel recommended that the ACHP not approve the draft Research Design — and it didn't. But the amended MOA allowed GSA to proceed without the approval of ACHP or that of the National Park Service, which also objected to Blakey's proposal.

Blakey, meanwhile, had gone back to the drawing board. But it should not be inferred that he was losing the battle with MFAT. For one thing, little if any word of his difficulties had percolated into the press. Second, he and his growing cadre of black supporters continued to stage angry rallies and public meetings to lambaste MFAT and "Euro-American" scientists. The tone grew uglier and uglier.

One GSA consultant, archeologist Nancy Demyttenaere, MA, of Saratoga Springs, New York, attended a meeting that included Blakey and MFAT members. In a letter dated January 25, 1993, Demyttenaere wrote to Federal Steering Committee chairman Howard Dodson, MA, saying that as a specialist in the preservation of recovered archeological material, she'd become "increasingly concerned" about the handling of the ABG remains. "On the basis of what I witnessed," she said, "my concern for the motivation and professionalism in this project turned to alarm."

Her initial concern, she emphasized, had been that the remains *not* be moved before being cleaned and subjected to stabilization treatments. But, "based on Friday's meeting, it is also now my professional opinion that certain behaviors and attitudes exhibited by... Michael Blakey actively threaten the preservation of these remains and their subsequent availability to unprejudiced scientific analysis."

Archeologist Demyttenaere continues:

> It is no secret that a certain level of academic conflict has evolved between Dr. Blakey and MFAT, but I was not prepared for the blatant animosity and power choreography of his attempts [to] remain isolated as the dominant scientific authority over the remains. His repeated incendiary baiting of MFAT members with statements of organizational ineptness and [his] slanderous claims of cultural, racial, and academic/scientific bias were wholly inappropriate to these proceedings, and I now consider them contributory to the dysfunctional progression of this project to date.
>
> Never before have I had the unfortunate experience to

witness in a public setting such inappropriate behavior by a professional and a scientist, let alone [by] an academician from such a respected institution.... As a scientist and an educator myself, I have very strong reservations about the motivation now driving this project.

None of the criticism had any discernible influence on the Steering Committee or GSA. The reason was that before GSA's deadline for comments on the draft Research Design, the Steering Committee had already recommended — on December 22nd — that the bones be sent to Blakey, at Howard, in Washington.

Judging by his colleagues' critical comments, as replicated and bound together by GSA, the committee failed to examine one potentially revelatory document about Michael Blakey and his scientific ability: There is no citation in these comments for, or discussion of, his major scientific enterprise thus far, the doctoral thesis in anthropology that he had completed at UMass in 1985. This document was and is readily available to the public from a commercial provider, University Microfilms International, in Ann Arbor, Michigan, or on the web at *http://web.archive.org/web/20010201063900/http://www.scholarworks.com/*

The purpose of doctoral research and a thesis is to demonstrate the candidate's ability to conduct a professionally qualified piece of original scientific investigation. UMass stipulates in its Anthropology Department Admissions Guide: "The doctorate in anthropology represents the specialized final degree to qualify an individual for teaching and/or research at the university level as a fully qualified professional anthropologist."

Blakey's thesis has the impressive title "Stress, Social Inequality, and Culture Change: An Anthropological Approach to Human Psychophysiology." In it he explores the destructive effects of racism on black immigrants in England. A salient feature is that this thesis is long on comments and opinion but short on analysis.

Specifically, the thesis explores the social, cultural, and personal effects of white racism on Jamaican and other West Indian British citizens who emigrated to England in the last century. Their hopes in coming — for security and betterment — have been frustrated. They find instead, Blakey says, only "shit work" and racial enmity from white Englishmen. They are isolated and stressed, to the detriment of their health.

Blakey is particularly impressed by the younger, second generation of these immigrants: They have embraced *Rastafarianism*, the nativist Jamaican movement that advocates separation from the dominant (here, English) society and the pursuit of one's own, indigenously black roots. "The Rastafarian [are a] relatively young section of [the] black community who engage a revitalistic cultural alternative to mainstream and British culture," he writes. Ironically, however, Blakey could not include these Rastafarians in his research study because they refused to respond to his western, scientifically constructed questionnaire: "What can these little ticks and boxes reveal of our condition and needs?" he quotes them as saying.

Their refusal skewed his research sample. But if some blacks didn't trust his white scientific ways, so, too, he says, many whites didn't trust him because he is black. "As an Afro-American researcher," he writes, "I could not expect honest answers from whites about their ethnic prejudices." This, too, skewed his study. Blakey does not indicate how he controlled his study to reduce such biases, other than by not reporting the whites' anti-black attitudes. He transcribed and included several long quotations from black subjects, but none from whites.

Partly because of these problems, Blakey had trouble recruiting subjects, whom he enrolled by "random[ly]" knocking on doors in Brent, a borough in London. He handed out 130 questionnaires. When he returned later, he managed to retrieve only half. So, he ended up with 65 respondents, of whom 21 were West Indian and 19 were English. The others were Irish, Asian, or another ethnicity.

Using note pad and tape recorder, Blakey obtained journalistically interesting interview comments from seven or eight blacks, only two of whom are identified as males. No English or Irish people are quoted.

Blakey's scientific findings, principally from his questionnaires, fill 16 pages of the thesis — a section called "Brent Survey" — of which 11 pages are text, and 5 are tables. These scientific findings constitute 6% of his thesis (16 out of 273 pages).

He calls it a "pilot study."

"The purpose," he writes, was *to begin* to test whether or not there is a substantial sense of social hopelessness in the working class, particularly the black working class, and whether measurable [effects of stress] are associated with perceptions of social and economic control [emphasis added]."

In other words, Blakey proposes, in the 1980s, "to begin" to test the basic proposition that received worldwide approbation, among communists and noncommunists alike, when Karl Marx and Frederick Engels first proclaimed it in 1848, in *The Communist Manifesto*.

Blakey's analysis of the data is bewildering, if for no other reason than it does not seem to lead anywhere. Blakey draws no conclusions from his Brent data, other than to assert that

SANKOFA?

there is an "association" between self-esteem and stress. There is no research news there! How many other psychologists, sociologists, anthropologists, and others had already reached this unsurprising finding? In fact, at one point, Blakey acknowledges, accurately, that "no conclusion is attempted in this report."

This lay reader finds the Brent Survey inept and confusing. Blakey's anthropologist colleagues might have found it more edifying. If it fills but a fraction of Blakey's doctoral pages, then one may ask, What fills the rest?

Many pages include discussions of theoretical matters, such as "characteristics of psycho-physiological stimuli," "the ontogeny of stress and coping," and "stress and cultural change." There are exegeses of other researchers' studies of culture, "race," stress, and related subjects.

None of it holds together. These many pages resemble stones, skipped across the waves atop deep oceans of scientific study and understanding. There are summaries, explanations, and commentary, none of which ever arrives convincingly at any shore.

It appears that Blakey read hundreds of scientific reports and other documents and noted information from each. Then, perhaps overwhelmed by it all, he strung his notes loosely together. If there is a meaningful trend, it eludes this reader. The document is full of sound and fury, but in the end seems to signify not much at all.

Blakey acknowledges being greatly "indebted" to others for help in his research and writing. "The [Brent] study would have come to little," he writes, "were it not for the tremendous aid" of two female and two male associates. He thanks 13 others for their help.

Blakey's Brent Survey and the rest of his thesis are not

simply a mishmash. But it is puzzling that the thesis fulfilled the doctoral requirements of a respected department at a major research university. Blakey's departmental chairman, Sylvia H. Foreman, PhD, and the five other members of his thesis committee "approved it as to style and content." (They are R. Brooke Thomas, PhD, Johnnetta B. Cole, PhD, George J. Armelagos, PhD, Alan C. Swedlund, PhD, and George W. Wade, PhD.) Blakey specifically thanks Cole, in the Acknowledgments, for "her careful attention to the manuscripts."

It is unclear whether any of the participants, pro or con, in the fight over Blakey's appointment to run the ABG research consulted this thesis. And, if they did, it is unclear what in it might have persuaded them that he was qualified to lead this important, costly, and highly visible scientific venture.

MFAT remained loath to let the bones go, albeit they had lain largely unattended in new steel file cabinets in the Bronx for several months. Without an approved Research Design, MFAT and Lehman College felt constrained to cancel their originally mandated plan, under Historic Conservation and Interpretation, Inc., to clean and conserve them. "Scientists at Lehman are prohibited from beginning the study at this time due to the lack of approval from GSA," the college's president said.

Blakey continued to excoriate MFAT for its "racism" because its eight members included no blacks. Turkel said later, in a phone interview, that he'd picked the best people he could find at that time.

Blakey suggested that the absence of blacks was deliberate. Certainly, at Howard, more black technicians and other research workers might be found. But Turkel pointed out that New York City had *many* more minority graduate students in the relevant disciplines, who were available for this hands-on work and the educational experience it would provide, than did Washington, DC.

Objectively, there were very few doctoral-level American black scientists qualified to head up such a project: perhaps three in all of anthropology, including Blakey himself and his close associate from the UMass at Amherst, Lesley Rankin-Hill, PhD.

By February, activists allied with Blakey were talking of using the legal doctrine of eminent domain to seize Lehman College buildings and the remains. They threatened a public protest that would damage Lehman's reputation.

"We could take huge amounts of the public down there, [and] raise the consciousness of people!" one activist, Howard Wright, declared.

In a letter to the Lehman president, New York State Senator David Paterson had warned: "[I]t is not incomprehensible that the affected [black] community, seeing no end to the mismanagement by MFAT, might well take matters into their own hands."

The GSA tried to hide the bind it was in. As part of its promotional effort, early in 1992 it had commissioned a film-making team, producer David Kurtz and writer Christopher Moore, to make a documentary, tentatively entitled "African Burial Ground: An American Discovery." Moore subsequently interviewed GSA administrator William Diamond in his office. Diamond said, on tape: "Are we in a powder keg situation? And the answer is, yes. We are. This is sensitive. It has been told to me, in no uncertain terms, that this could be a, quote, 'Rodney King situation.'"

The late Rodney King was a black motorist in Los Angeles who, in 1991, was chased by the police, who beat him up, unaware that a nearby observer was taping them with his minicam. When this brutal scene was played, repeatedly, on TV, black riots ensued.

GSA's Diamond told Moore that this example "means that I have to move, and we [GSA] have to move with sensitivity, understanding, and that we have to be as flexible as we possibly can be."

"But," he continued, "at a certain point there is no flexibility.... [T]he government will not be ... blackmailed or blackjacked or ... threatened into a position that would be contrary to law.... And thank God we have not reached that. I hope we never [do]."

Later, just before the film was to be previewed for the Steering Committee and other ABG stakeholders at Trinity Church, GSA cut out this segment. Moore protested, saying the deletions "severely compromised" the film's integrity. Much later, in a phone interview, he claimed that the deletions were restored. A screening of the film at the ABG library, in Manhattan, confirmed that they largely were: Only the last two sentences (above) were omitted.

All in all, by spring 1993, Blakey was ahead of the game, despite all the criticism. It might have been different if the Steering Committee or GSA had had scientist members who could articulate Blakey's deficits in chambers or if colleagues in the community would speak out *publicly* — but neither happened. And MFAT had no help from the press; Blakey's was the better sounding story.

Blakey was enjoying significant success in dealing with GSA officials. Partly, this reflected his skill at playing the race card. "GSA's top management didn't question him," explained later Peter Sneed, the GSA urban planner who was in charge of the Fed's 290 Broadway venture. Sneed says, "A lot of uptight white people were afraid of him. He played that

SANKOFA?

[fear] like a musical instrument."

The regulatory process that protected the remains and would pay for the research meanwhile moved ahead. The ABG was designated a New York landmark in February 1993. In April it became a National Historic Landmark.

In a major development during the spring, an Office of Public Education and Interpretation (OPEI) opened its doors at 6 World Trade Center, not far from the ABG, under a government contract with John Milner Associates (JMA). The mitigation efforts included, by law, a major effort to publicize the site and convey its historic importance to students and teachers, community leaders, and the public at large. OPEI's chief was an urban archeologist, Sherrill D. Wilson, PhD, who became, after Blakey, the best-known and only other ABG spokesperson.

A forceful black advocate, Wilson was dismayed by the racism she found among white archeologists, the profession to which she aspired. After two years quizzing classmates and faculty at New York City's New School for Social Research concerning what archaeology was all about, she said she concluded that it was "the study of non-white people by white people" — a situation she set out to change.

With a generous annual budget from GSA via JMA, she was to develop OPEI into a potent social advocate for the research and the historical importance of the ABG for black Americans. A major tool in this effort was a quarterly newsletter, called *Update*, with Wilson as top editor; it was mailed, free, to thousands of individual and institutional readers.* OPEI's "general goals," *Update* reported in its initial

---

* GSA paid for this work under a contract with the JMA. Through April 2009, GSA says, it paid the company $7,482,377.35, much of it for OPEI-related activities.

(Spring, 1993) issue, included "educating the public on the lives and neglected histories of the population . . . of the ABG."

OPEI's main messages through the years were the promotion of black uplift and black pride. The publication and OPEI often opposed the "white" US government — a peculiar position, given that OPEI workers' salaries were paid by the feds.

There was another, burning imperative for OPEI. Education. For blacks, for whites, for everyone: To tell what was being found and what it meant. Particularly, to depict the horrors of slavery as deciphered from the human remains and archeological and historical evidence. Particularly, to show that slave labor was (allegedly) a major force in building New York City and that it was (allegedly) practiced on a scale and with cruelty comparable to the South. In their suffering and in their numbers, Blakey declared, Manhattan was the birthplace of the America's African American people.*

---

* But: Ninety percent of slaves in America lived in the South at the end of the 18th century. By 1810, there were a million slaves there, compared to a few thousand in New York City.

SANKOFA?

7

SCIENCE AND INSPIRATION

MICHAEL BLAKEY RETURNED TO the drawing board to revise his scientifically criticized draft Research Design. Meanwhile, as prescribed by the National Historic Preservation Act, the General Services Administration (GSA) set up ad hoc committees of his scientific peers to evaluate the plans. GSA sought to "determine the proper treatment of the human remains," it told Howard Dodson, the Steering Committee chairman, by letter. Two committees, with some members serving on both, were appointed. One would review Blakey's plan to stabilize the remains and ship them to Howard University in Washington. The second, more important panel was tasked to "review the Research Design, as revised."

The reviewers, Dodson was told, were chosen on the advice of the scientific community, on his Steering Committee's recommendation. All were doctoral-level specialists, with subspecialties in the study of old bones and burial sites. Some panelists were black, some white. They were the late Philip L. Walker, PhD, of the University of California, Santa Barbara; Theodore (Ted) A. Rathbun, PhD, at the University of South Carolina; Clark S. Larsen, PhD, later at Ohio State University; Carrel Cowen-Ricks, who was working on her doctoral dissertation on early African American cemeteries and graveyards; Eleanor Mason Ramsey, PhD, who later started an archeological business in Oakland, California; and Jerry Rose, of the University of Arkansas.

This panel was extremely receptive to Blakey and his plans.

Anthropologist Larsen already was on record approving Blakey's research plans as "fully adequate." Three others had disinterred and studied remains from an historic black cemetery — Cowen-Ricks in coastal Georgia, Rathbun a slave cemetery on a South Carolina plantation, and Rose a Reconstruction-era church cemetery at Cedar Grove in Arkansas.

No Metropolitan Forensic Anthropology Team (MFAT) members or other New Yorkers were appointed to this panel. MFAT co-director Spencer Turkel, PhD, later would allege, in a telephone interview, that Blakey had chosen these peer reviewers — a possible problem, if true, because scientific reviewers are supposed to be independent commentators.

Blakey gave GSA his second revised draft Research Design in April. Copies were sent to the reviewers. It clarified, somewhat, his plan to use genetic data to determine cemetery denizens' ethnic origins: By using specific genetic techniques (to be described below in Chapter 18), he hoped to obtain genetic sequences — profiles — from individual sets of remains that could be compared with similar sequences from individuals in present-day African and other populations. Matches between the African Burial Ground (ABG) DNA and DNA

SANKOFA?

from people from known contemporary groups would strongly indicate whether the remains were African or not. The only problem was that few genetic studies had been conducted on people in southwestern and south central Africa, where, based on historical information, most enslaved Africans had originated. Blakey pledged to take the ABG project to Africa to obtain this genetic data.

He then could cross-check this DNA finding with data from analyses of the ABG population's bones and teeth and dietary information from dental calculus to establish an individual's provenance. These interred people then could be grouped as Africans. Others, who failed these tests, could be excluded as having non-African origins, e.g., European or Native American.

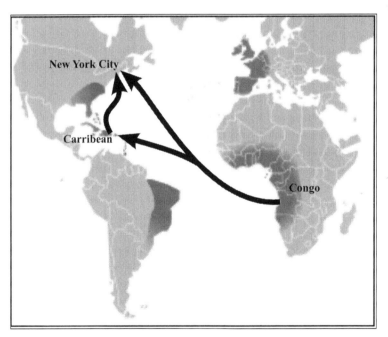

Middle Passage Splits. Slaves in New York were shipped from Southwest Africa. Some came directly. For others, they or their families were taken first to the West Indies. Between 1701 and 1726, 605 came directly from Africa and 1,348 from the West Indies, according to the *ABG Archeology Final Report.*

MEANWHILE, ONE KEY BLACK voice spoke up against Blakey's plans. New York City's Mayor, David N. Dinkins, opposed removal of the bones and research to Washington. In a May 21, 1993, letter to Congressman Charles Rangel (Dem., New York), he expressed "serious concerns" with the Steering Committee's recommendation that the project be moved from Lehman to Howard.

"I am extremely distressed that this project would leave New York City," Dinkins wrote. The ABG "is part of New York history and it belongs here," he added. "In addition to the loss of cultural and intellectual stimulus for NY residents, professionals, students and institutions . . . the relocation of this project also means the potential loss of substantial research dollars."

For these reasons, the mayor concluded: "I cannot support a resolution which moves the whole project out of New York."

A rebuttal, favoring Howard and Blakey over MFAT and New York scientists, reached the mayor's desk within a week. It came from Steering Committee chairman Dodson. The reasons for recommending the transfer, he wrote, reflect "the opinion of a broad spectrum of the African-American Community that [the] research . . . should be done within the context of an African-American institution . . . under the leadership of African-American scientists/scholars."

Why?

"European and Euro-American scholars have a long-standing history of misinterpreting the African-American experience. It was felt [by the Steering Committee] that the level of cultural sensitivity . . . required could best be achieved within the context of an African-American institution . . . . This position was expressed forcefully by diverse segments of the African-American Community as soon as it was determined that the initial exhumation activities were being carried out by a field research team that had no African-Americans in its leadership structure and few if any were working on the project."

MFAT, Dodson said, has shown on "innumerable occasions" that it is not the appropriate agency to manage research activities associated with the interpretation of this "invaluable resource."

Dodson charged that objections to Howard and Blakey had been raised, first, by MFAT "to discredit Dr. Blakey and to try to position itself to be a major participant, if not *the* principal research team" [emphasis in the original].

In reaching its decision, Dodson noted, the Steering Committee members had reviewed "competently written draft research design proposals by Dr. Blakey, and found them professionally superior and much more historically and culturally sensitive than work done by the previous research teams." He added:

"We have been motivated primarily by a desire to ensure that the research activity will produce scientifically credible and culturally and historically sensitive results."

The design proposals that Dodson and his committee had "found professionally superior" to the ones prepared a year earlier by Edward S. Rutsch of Historic Conservation and Interpretation, Inc. were, of course, the ones that some commentators had lambasted a few months earlier — and which Blakey had just revised. Nevertheless, in concurrence with the Black Community's wishes, Dodson said, he and his committee members had placed themselves solidly behind Michael Blakey.

No vote, it should be noted, was ever taken among blacks in New York City, or elsewhere, to decide how the ABG remains should be dealt with and by whom. While many black and white New Yorkers were appalled by GSA's high-handedness, the number of activists dwindled over time, until their mandate derived from a small cadre. Many, moreover, eventually turned against the project because it was delaying the respectful reburial of the remains — which was their principal concern. Ironically, Blakey would turn out to be responsible for much of this delay.

Blakey's relationship with the Descendant Community reflects what one preeminent black social analyst has called "an almost incurable Messiah complex, characteristic of Negro emotionalism." In his book *The Crisis of the Negro Intellectual* (William Morrow, 1967, N.Y. Review of Books, 2005), Harold Cruse explains:

"There must always be the great Individual Leader — the Messiah, the Grand Deliverer, the cult of the Irreproachable Personality who, even if he does not really have all the answers to the problem, can never be wrong. Negroes . . . are led by their emotions rather than by reason or the guideposts of social analysis. This [is] political immaturity . . . ."

After Blakey submitted his new draft, GSA convened a meeting of the Research Design panel at its headquarters in Washington. They discussed the new draft. On May 14, 1993, the panelists visited Blakey at Howard, spoke with university administrators, and toured the lab. On their last day in town, they returned to GSA to conclude their discussions and write their report, which was submitted to GSA and sent to Dodson's Steering Committee shortly thereafter.

The peer reviewers' final report clearly notes continuing conflict between Blakey's research plans and standard scientific methods. They say the draft is "much improved" over earlier versions — but they still found it inadequate.

SANKOFA?

Assessing their visit to Howard, they say they were "very much impressed" by the university's "capability and commitment to the project." With additional changes, per their present recommendations, they had "little doubt" that Howard "can carry the project to a successful conclusion."

Then, following this optimistic pronouncement, the panel presented a 19-page, single-spaced critique of Howard's resources and Blakey's research design. The university is "weak in historical archaeology," according to the critique, and the design may have "underrat[ed] the project's need for centralized administrative support." Howard, the panel believed, needed greater control over the work.

The panelists were "very much concerned" by the proposed separation of the project's three parts, in which Howard would do only the bioanthropology, while the archeological and historical research and writing would go elsewhere:

"Such a division would almost certainly guarantee inefficiencies, loss of data, and inadequate ... research and ... public information." The entire project, the panel reiterated, should be carried out under one leader's direction, at one university — preferably Howard.

Blakey's plan for biohistorical research on the remains seems "overblown," the peer reviewers added. "The focus of this project should be on research that realistically can be accomplished."

The *soil pedestals* — the earth beneath the graves into which the bodies had decomposed — are "the best and most likely source" of DNA that has *not* been contaminated with DNA shed from the bodies of technicians and others who had unearthed them, the panel said. So the panel urged great care in removing the remains and underlying soil and stipulated that GSA should not pay for major DNA analyses "until and unless the likely presence of uncontaminated DNA is demonstrated."

The panel said Blakey and his associates should focus on *basic* research and on developing an overall view of what might be learned from later, more specific research studies. It disagreed with his plan to focus the research on "a narrowly defined time period in the 18th century, rather than on the entire history of the site from the 17th to the 20th century."

Photo provided. Michael Blakey gave this image of himself to *Current Biography,* which published it with his profile in 2000. Profile described his career.

SANKOFA?

THE RESEARCH DESIGN IS "rather simplistic" in its proposals for studying the cultural influences that shaped the cemetery's burial practices, the panelists continued. The design reflects "inadequate attention to, and probably [inadequate] knowledge of, the relevant literature" on African, Muslim, and other burial traditions.

The panel also listed many specific aspects in the design that it felt were either "inaccurate or required [further] discussion." For example, the Design "fails to address what is known about Congo and Angolan burial traditions" that slaves might have brought with them from Africa. Similarly, it "fails to justify using" the remains to study environmental toxicology in America.

Blakey proposed to study the remains in light of a number of factors, including the black churches. The panelists note that the historical literature "suggests there was no African church as such at this time . . . ." The design also envisions studying ship manifests and other documents in Europe. The reviewers note that much of the documentation of the slave trade had been copied and is available in the United States. So foreign travel to find the originals, as Blakey proposed, is probably not needed.

Proposed comparative studies with African populations could require "very expensive" overseas research. These plans require more "specification and justification."

The Research Design proposal should, but does not, address the question of "What happened to the descendants of the people buried in the ABG?"

The document overvalues the burial ground population: It is "an" African American population, not "the" African American population. "Without denigrating [its] importance . . . it should not be assumed that the ABG is central to the historical and cultural integrity of all African-Americans everywhere."

The "feasibility and efficacy" of the proposed chemical and DNA studies "are questionable," according to the panel. Blakey proposed to analyze the bones to determine whether the people grew up eating corn or millet, as an indication of whether they were raised in America (on corn) or in Africa (on millet). But this can't yet be done, the panelists said.

In preserving the remains for future reburial, the panel found no justification for Blakey's proposal to avoid ultraviolet light exposure. Similarly, wrapping the bones in acid-free paper — as Blakey had insisted, in chastising the MFAT disinterments — was *not* necessary (as discussed above in Chapter 3).

Despite their many general and specific objections to the revised design draft, the panelists nevertheless said that they did not want Blakey to go through yet another set of revisions. Rather, they suggested that a new document, specifying the scope of work for the first part of the project, be developed henceforth by GSA and Howard, based on the revised Research Design and this, the panelists' final review of it.

Still, doubts remained: One panelist jotted in pen on the title page of his copy of the revision:

- "Due to the size and expense of this project, the participants will be under great pressure to do a job that will stand up under peer review."
- The Research Design "never really says what data will be collected from skeletons."
- "There are value statements concerning neglect of African-Am[erican]s that are patently false."

Why the panel or GSA failed to resolve the many problems raised in the panel's report is unclear. GSA was then under intensifying pressure from activists in the Black Community to move the bones to Howard and get going on the research so that the bones could be quickly returned to New York for reburial. The GSA regional officials in Manhattan, who were overseeing the new building and, now, the ABG, undoubtedly were eager to get the bones out of town to reduce the threat of violence at their doorstep.

It was quite exceptional for GSA to fund a major scientific project. Unfortunately, it lacked the institutional structure and knowledge to do so. GSA dispenses money through *contracts*, not through grants. Judged by the negative and unresolved problems raised by its consultants about Blakey's Research Design, it is clear that federal *granting agencies* for science, such as the National Science Foundation and the National Institutes of Health, would never have approved or paid for Blakey's plan.

Despite the doubts, however, Rose said, a decade later:

"We — the Committee — were all optimistic!"

David Zimmerman

8

Moving Bodies

By August 1993, the Federal Steering Committee had received, assessed, and approved the Research Design panel's report, and had forwarded it to the General Services Administration (GSA) and Congress. On August 12th, GSA signed an agreement with Michael L, Blakey to remove the skeletal remains from storage at Lehman College and have them shipped to Howard University. Could it be done without damaging the fragile remains? Who was to do it?

Black people "should . . . participate," one community member, Gary Williams, had insisted at a recent public forum. The company "that is going to transport our ancestors' remains," he said, should have African Americans or Africans in it. "Money [is] involved, definitely, and I think we should also benefit from that." Blakey demanded instead that the moving company be one that had expertise in moving fragile human remains. At his recommendation, Artex Fine Arts

SANKOFA?

Services, a nationwide company, was chosen.

In September, he received a letter from a GSA official, Lydia Ortiz, objecting to the "ethnocentric overtones" in the Research Design. "Please understand," Ortiz wrote, "that the U.S. government may not be a party to, or engage in, any form of discrimination, either in acts or language. Accordingly, please review the entire Research Design, deleting any discriminatory references, inferences, or attributions . . ."

Blakey shrugged off the GSA complaint, saying, much later[*]:

"In fact, no changes would be made, because no discriminatory content existed."

Moving the bones was costly. But the spigot now was open, and federal funds had begun flowing. In August, Howard University signed a contract for "packing, moving. . . transportation of the remains for a cost of $261,481," or about $650 per set of remains, each of which, on average, consisted of three file drawers full of bones, earth, and other material.

Artex trucked the first shipment of 13 sets of remains to Washington in mid-September. The researchers carefully examined them and found no damage. Blakey then had the remaining 400 or so sets of remains sent to Howard, heralded by a candlelight vigil at the African Burial Ground (ABG) site and a celebration called "The Ties That Bind" upon their arrival in Washington. GSA paid an additional charge of about $6,000 for videotaping these ceremonies.

Officials in GSA's New York regional office may have heaved a huge sigh of relief once the bones were safely out of town. Members of the Metropolitan Forensic Anthropology Team (MFAT) continued to stew.

---

[*] This racial selectivity is part of the documentary record. A peer review panel had declared: "All scientific . . . research should be directed by African-American professionals . . . ."

Removal of the remains to Washington remained controversial, even among people involved in or close to the ABG project. In August, a writer, Rodger Taylor, interviewed John Milner Associates (JMA) staffers and members of the public on the move:

"I have no objections about them going to Howard," declared Esther Dawson, whom Taylor identifies as a concerned citizen. "Because," she said, "these are black scientists. What could be better than to have our own people study them!

"I also think it's important that they share the information with us in New York."

She added:

"It would also be interesting . . . if the scientists would write and tell us if they got any spiritual feelings from working with the remains."

Claudia Milne, a graduate student who was then working on the bones at Lehman as a lab technician, said that from her perspective, "It's a loss for New York City and . . . its students. But if it's been determined by the Community that the people at Howard are the best to do the job, so be it! I hope it goes well. It seems like there is a possibility that more damage will be done to the remains by transporting them."

"Howard doesn't have a graduate department in archaeology," Milne noted. "Graduate students usually write the papers that inform the greater community . . . . Are they [Howard] going to build an archeological department based on this project? . . . . I'm curious about it."

Another lab technician said that on the one hand, it was "somewhat unfair" to New York City students to remove the remains to Washington. On the other hand, Doville Nelson wondered whether the MFAT researchers would have been "competent and sensitive enough to do an effective job."

SANKOFA?

"I don't think so."

"Europeans [white people] tend to see our remains as a data pool, to be objectified and in that way minimalized ... whereas black people see these remains ... as a life experience and part of the continuum of our experience."

Historian Christopher Moore, a Steering Committee member, said, "I don't think where the remains go is as important as making sure a qualified African-American is head of the project. At the moment, that person is Dr. Blakey."

As an artist and activist, Miriam Francis, also a Steering Committee member, said after the move was complete:

"I fully respect Dr. Blakey, and ... wanted the remains to go to Howard. So I'm satisfied that they are. My understanding is that one of the things that Dr. Blakey and his colleagues will ... tell us is what part of Africa our ancestors came from. Learning that would be satisfying to me."

Place of bondage. Slave market stood by East River at end of Wall Street.

MFAT MEMBERS, WHOM WRITER Taylor did not quote in his article, of course begged to differ. Anthropologist Leslie Eisenberg said later that she and the other MFAT workers viewed the remains objectively. "We tried not to have an agenda," she explained. "We were collecting data in the field, fully expecting to do detailed examinations at Lehman. We had not made any interpretations at all."

Like her MFAT colleagues, Eisenberg objected to sending the remains and the research to Blakey and Howard.

"When you have qualified people in the field [here in New York], why would you call Washington, just because someone there was black?" she asked.

Besides, "I thought it was wrong to take remains from the city out of the City!"

Meanwhile, new and final revisions for the Research Design were worked out among Blakey/Howard, JMA, the Steering Committee, and GSA and were published by GSA on December 14, 1993.

Two stipulations to direct the presentation of the site's "extraordinary scientific potential and public significance" are spelled out in an amendment: "Presentations and professional meetings and publication in scholarly journals are the media by which researchers generally disseminate data to the wider scientific community." This last draft states that these scientific forums also must provide feedback to foster other researchers' efforts:

"The ABG data, as a scientific resource of exceptional significance, needs to be made available to the professional community *as work progresses*, and prior to reburial and publication of a final technical report" [emphasis added].

The amendment also stipulates who is to do this reporting:

SANKOFA?

First, only researchers who are "formally involved" in the ABG will have access to the data while the research is in progress, an estimated six years. Second, "the Project Director" — Blakey — "will act as senior editor and author of final technical reports." In other words, Blakey had won control of all the scientific findings that would be forthcoming while the research was in progress.

In the final amendment to the Research Design, Blakey wrote this succinct and fairly clear statement of his research goals:

> *To strengthen the analysis ... of population affinities ... sufficiently, correspondence between probable genetic affinities will be sought with anthropometric similarity and dietary/environment proximity to the hypothetical parental populations. When DNA-based genetics, anatomical structure, chemical signatures for environments, and cultural traits conform to those of probable parental populations, the origins of individuals comprising the African Burial Ground population can be ascertained.*

The challenge was that the first of these overlapping data sets, the skeletons' anatomical structure, was based partly on intuition, not data: Blakey indicated that he could identify black — African or African American — human remains simply by looking at them — or by "eye-balling" them, as one critic had said. The only data Blakey would consider in this respect would come from the skull, particularly the teeth; he claimed special expertise in dental analysis. Most other "racial" analyses of old bones relied heavily on measurements and comparisons of bones *below* the skull. These analyses were grounded in actual measurements of buried skeletal remains of individuals of known ethnic origins.

The possible use of DNA genetics was similarly uncertain: It depended on the possibility of obtaining viable DNA from the remains. And, since there were, and still are, few reliable genetic studies of populations in the African region where most enslaved blacks were believed to have originated, Blakey's proposed analysis would be dependent on these studies being done — a costly and difficult task. Populations would need to be studied from Senegal-Gambia, Berin-Congo, Khoi-San, Northwest, Southern, and East Africa, as well as Holland, Great Britain, and Native Americans in what is now the northeastern United States.

Finally, there was not then any scientifically accepted method for determining skeletons' place of origin by studying chemical signs of nutritional differences read from the bones.

Attempts to determine the remains' cultural origins also were, and are, dicey, since what was left in the graves, 200 or 300 years later, provided few specific clues to the individuals' cultural affinities.

All in all, Blakey's ABG Research Design remained tenuous.

Blakey won one other important victory. The document states that the Steering Committee's approval of the design "formally establishes public support for the standards of data recovery and dissemination" of the findings.

This community approval is required by the National Historic Preservation Act of 1966 (NHPA), under which the project was unfolding. Nevertheless, many Steering Committee members were interested parties who shared Blakey's point of view, albeit they were only a tiny minority of New York City's black population — among whom no vote on the ABG project had ever been taken.

Blakey later would claim that the committee's approval obligated GSA to fund *all* research described in the document, even when it cost more than GSA's budget and federal guidelines allowed. He later extended this to mean that all of the work he planned was mandated and preapproved under the provisions of the NHPA. Given the Research Design's imprecision, this granted Blakey a virtual *carte blanche* — or at least, that is what he would say later. MFAT members and others might demur.

DAVID ZIMMERMAN

9

WHAT MIGHT BE FOUND

WHERE MIGHT ALL THIS effort lead? What could the African Burial Ground (ABG) population tell scientists — and the rest of us?

These questions and their answers were of particular interest to a small but growing cadre of researchers, principally physical anthropologists, who were excavating graveyards, both ancient and recent, at sites throughout the Americas, and elsewhere as well.

Jerry Rose was one of them. Ted A. Rathbun of the University of South Carolina, who also had served on the Research Design Peer Review panel, was another. A third was University of Oklahoma anthropologist Lesley Rankin-Hill, a contemporary and colleague of Michael Blakey at UMass, who had focused her work on Afro-American biohistory. Blakey had worked with her on a study of skeletal remains from the First African Baptist Church in Philadelphia. Blakey planned to appoint her as his associate director for the ABG project.

What the skeletons had to offer science was *data*, i.e., observations and specific and detailed measurements of the exhumed bones. Observations and measurements that, for starters, indicated the stature — the height — of the individuals at the time they died. For adults, but not for infants or children, the pelvic bones, if they remained, indicate a person's sex; women have wider pelvic girdles, to facilitate childbirth, than do men. The teeth, jaw, and facial shape provide a strong, but less-than-perfect readout of ethnic identify: Native American, white, or black. Some Native Americans, for example, have what anthropologists describe as shovel-shaped incisors.

The skeletal anatomy might show much about its owner's health. Some bulged bones are a clear sign of healed fractures. Infectious diseases, such as syphilis and tuberculosis, and the vitamin D deficiency disease, *rickets*, leave telltale marks and deformities of the bones. Heavy use of joints and hard physical labor remain discernible in death, also leaving osteoarthritis in their wake.

A person's age at death can be discerned from his or her bones. A group of skeletons can suggest their owners' longevity and can be used to calculate the group's fertility rate, if a large enough number of burials have been unearthed.

Defects in dental enamel, called *hypoplasias*, reveal nutritional stress before birth and during childhood. Spongy overgrowth of bone in the orbits of the eyes and other sites show the presence of iron-deficiency anemia and sickle cell anemia, debilitating conditions that stunt body and mind. The bony overgrowth represents abnormal expansion of the bone marrow as it struggles to produce more red blood cells to relieve the anemia.

Could these and other scientific data from human remains provide a measure of the health and well-being of early Black

Americans? More specifically, could they provide a window to life under slavery? How healthy were enslaved Blacks in colonial and early American places? How good or how poor was their well-being?

These are not simple questions. Jerry Rose had written a few years earlier: "The history of Black Americans is a complex subject, which has engendered heated historical debate for more than a century. Much of the debate has been focused on the quality of Black life and existence, especially diet, mortality, and health."

He added: "[N]o other aspect of American history seems to generate so much interest, emotional involvement, and research as the institution of southern slavery."

The occasion for Rose's observation was the preparation of his report *Gone To A Better Land* (Arkansas Archeological Survey Research Series No. 25, 1985.) This is his study of the human remains in the Cedar Grove Cemetery, which he disinterred and studied very briefly before they were reinterred farther from the Red River, which had threatened to wash them away. Rose was one of the first researchers to excavate a black cemetery in America.

Cedar Grove had been used for burials from 1890 until 1927, when the river flooded and covered the cemetery with silt. Some of its inhabitants had been born slaves and died free. Others were born and lived during Reconstruction, and had toiled as sharecroppers under the harsh new relations between whites and Blacks that developed after the Civil War. Rose and his colleagues examined a number of sets of remains, as well as the coffins, clothing, and other physical material found with them.

But they were not ready from this preliminary analysis to draw firm conclusions from their findings.

Rose cites Kenneth Stampp's *The Peculiar Institution: Slav-*

SANKOFA?

ery in the Ante-Bellum South (New York: Vintage, 1956), in which Stampp had written that slavery was inhumane and slaves' living conditions were substandard. Subsequently, Robert Fogel and Stanley Engerman, in their massive *Time on the Cross: The Economics of American Negro Slavery* (Boston: Little Brown, 1974), using more sophisticated analytic methods than Stampp, argued that slaves' living conditions were no worse than whites' and their diets were nutritionally adequate. A major debate then ensued. Two authors, Kenneth F. Kiple and Virginia H. King, declared, in *Another Dimension to the Black Diaspora: Diet, Disease, and Racism* (Cambridge, 1981), that the slave diet would have been adequate for white people but not for blacks, because of biologic differences in their nutritional needs. Many blacks, for example, were lactose intolerant genetically, and so could not digest or thrive on dairy products. They shunned milk and thus were calcium deficient. Slaves' diets in the South were based on corn, which provides too little iron, leading to anemia.

These disagreements forced historians back to the data, and to the understanding that quality of life and diet varied from place to place. So, calculations of blacks' well-being under slavery would have to be focused on specific areas and sites.

"[W]hat better way to assess nutritional adequacy," Rose and a co-author say, "than to *directly* examine its impact on the human body by analysis of historic human skeletons[!]" [emphasis added]. Such studies, they add, "have provided valuable data on Black diseases, diet, and quality of life."

"Human bone," explains Blakey's UMass colleague, Rankin-Hill, "provides an excellent source of health status information because it is metabolically initiated, nutritionally tempered, physiologically controlled, and biomechanically shaped. Bone provides a measure of biological and cultural factors that have affected the health of an individual/population."

This, then, is a principal scientific rationale for studying, where available, black human remains.

By the mid-1980s, more than a decade after the Cedar Grove disinterment, economic historian Rose, and other like-minded researchers had made an impressive advance: They had assembled comparable data from dozens of colleagues on 12,520 individuals exhumed from 65 burial sites throughout the Western Hemisphere. Some sites were ancient, even prehistoric, while others were scattered through the centuries to recent times.

The majority of the grave sites were Native American. There were fewer Euro-American, and only five African American sites, including Cedar Grove in Arkansas and the black Baptist Church cemetery in Philadelphia. These sites accounted for only 1,380, or a little more than 10% of the individual sets of remains in the study. If the ABG remains from New York City were added, they would have increased the number of blacks in the sample by almost one-third. But Blakey missed the deadline with his data, which were published separately, according to Rose.

In return for the ABG information, the database that Rose, Steckel, and their associates were developing might offer the ABG researchers this benefit: an objective, albeit preliminary, evaluation of 17th- and 18th-century black New Yorkers' health and well-being, compared to the health and well-being of 65 other groups of ancient and later African American, Euro-American, and Native American peoples.*

---

* Much of the recent progress in understanding early humans and the evolutionary transitions from ape to man is based on comparisons between fossil teeth, and on comparisons between fragments of skulls and other bones.

SANKOFA?

MEASURING BONES AND COMPARING other measurements could produce hundreds of types of data — many too many to be useful. So the researchers winnowed the possible data sets to obtain a manageable number of measurements — seven — that could be obtained from a large number of remains and might reveal the most about their owners' health:

- The length of the thigh bone, the femur, indicates a person's overall height in life. The taller one is, in general, the healthier. Short femurs, indicative of short stature, suggest nutritional deficiencies or other forms of childhood and adolescent ill health.
- Defects in the enamel of baby teeth and adult teeth are "an indelible indication of periods of stress during tooth development" from before birth to age seven according to two physical anthropologists from Hampshire College in Amherst, Massachusetts, Alan H. Goodman, PhD, and Debra L. Martin, PhD. These linear enamel hypoplasias are marks of ill health and, later, functional impairments.
- *Porotic hyperostosis* (abnormal thickening of bone tissue) also is scored as a revelatory marker of health and well-being. Even when it is mild, iron-deficiency anemia has a profoundly negative effect on health. It lowers one's resistance to disease, reduces the ability to work and play, and can lead to mental deficiency. Goodman and Martin say IQ tests "suggest that iron is a critical element for the normal functioning of the nervous system, and that cognitive functions can be disrupted by relatively mild iron deficiency." Severe deficiency may be worse. The detection of porotic hyperostosis in a long-buried skull suggests that its

owner suffered from ill health and functional impairments.

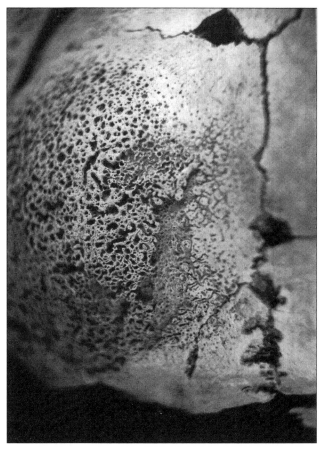

Bone distortion (porotic hyperostosis) on normally smoothish skull surface is hallmark of anemia due to dietary iron deficiency, sickle-cell disease, or both. Body creates extra bone to make red blood cells to carry iron to all its tissues.

- An individual's dental health at the time of death is also accessible. The number of teeth and the number of cavities (caries) can be counted. The more teeth that remain and the better their condition, the better the owner's health.

SANKOFA?

- The long bones of the leg, the tibias, are examined for signs of disease, several of which, including infections by staphylococcal or streptococcal bacteria — "staph" or "strep" as they are commonly called — leave marks and scars in their wake. The leg bones and any other available bones also are examined for signs of tuberculosis, syphilis, and other infectious diseases.
- Degenerative changes in the bones, such as osteoarthritis and rheumatoid arthritis, are also counted. These defects reflect chronic physical stress on the joints and thus may indicate an individual's activities in life. Running, for example, stresses the knees; rowing a boat affects joints in the elbows and shoulders. These pathologies can be quite painful and detract from the quality of life: The more such lesions are found, the less healthy their owners are deduced to have been.
- Trauma, including broken bones from accidental falls and injuries from violent confrontations with animals or other humans, leaves permanent marks. Injuries to the arms, legs, nasal bones, face, skull vault, and hands are scored, as well as weapons wounds. The fewer wounds a person sustained, the better his or her health is judged by anthropologists to have been.

Now, how do we make sense of all these data?

Anthropologist Rose and two colleagues — economist Steckel, and anthropologist Paul W. Sciulli, PhD, also of Ohio State University — entered the findings from the 12,520 skeletons into a database. They called it the Mark I version of their computer program for determining dead persons' Health Index. And, more importantly, they programmed it to derive average numerical scores for each of the 65 groups of burials in which these individuals belonged. They then listed the 65, starting with the healthiest groups on top, descending to the least healthy group on the bottom. This provided a ranking of what they called *quality-adjusted life-years.*

The comparable health of groups of individuals from widely divergent times and burial sites could now be determined. This was a major scientific advance.

The best score — the highest quality-adjusted life-years — belonged to a group of Native Americans who lived along the Atlantic coast in what is now Brazil some 1,200 years ago. The lowest scores — the poorest health — was recorded for Native Americans who lived in what is now New Mexico some 450 years ago, around the time of the Spanish conquest.

Of the five African American sites, two were near the bottom in terms of quality-adjusted life-years: denizens of the plantation slave cemetery in South Carolina, studied by Rathbun, and those from Cedar Grove. Only one group of African Americans fell in the upper half of the 65 sites in terms of health, the denizens of Philadelphia's 19th- century Baptist Church cemetery. Their health and well-being was *better* than that of occupants in eight of the nine Euro-American (i.e., white) cemeteries in the study.

The obvious question is: Where would the 400 ABG remains fall on this list?

GATHERING, ANALYZING, AND EDITING these data for publication took several years. The Mark I study and supporting scientific papers were not published until 2002, in a volume called *The Backbone of History: Health and Nutrition in the Western Hemisphere* (Cambridge University Press). The editors, Steckel and Rose, state explicitly that theirs is a pioneering effort toward the creation of a universally applicable method to

SANKOFA?

evaluate a particular population's health and well-being based on its skeletal remains. Refinements will be forthcoming, they say. Nevertheless, this initial scheme provided a framework into which the ABG findings might informatively be fitted. The Health Index might provide objective answers to key questions raised by Blakey and others: How good was the health of these early Black Americans? What was the quality, generally speaking, of their lives?

To answer these questions, Blakey would have to collect data from the ABG remains that were prescribed in the guidelines for the Steckel, Scuilli, and Rose study, including "race." Disappointingly, he would fail to do so.

David Zimmerman

10

Digging In

A NEW PHASE OF the African Burial Ground (ABG) project opened in 1994. It was a clear break from the raucous, rancorous events in Manhattan in the previous three years: The time had come to get down to business.

Blakey's first task was to prepare his facilities. The Research Design panel had given Howard University high marks for its "capability and commitment," saying its resources are "more than adequate" to "carry the project to a successful conclusion." One panelist had described Howard's facilities as "first rate." In point of fact, Blakey's area, the Cobb Laboratory, consisted, quite modestly, of three laboratory rooms, a hallway, and two offices, for a total of 3,000 square feet (which is equivalent to a 55-foot by 55-foot space). At the outset, Cobb lab had only two work stations at which skeletons could be laid out, cleaned, repaired, and measured. A third was added. Howard's resident faculty was similarly limited in size.

SANKOFA?

Blakey was the only anthropologist. But he enlisted three other doctoral-level faculty members for the project: One, Matthew George, PhD, was a biochemist and molecular geneticist, which made him well qualified to perform genetic analyses of the cemetery's remains — provided, of course, that DNA could be extracted from them. The second Howard specialist was a surgeon, S.O.Y. Keita, M.D., who also had master's degrees in biology and anthropology. The third, Emory J. Tolbert, PhD, was a historian, whose C.V. revealed little interest in bioanthropological studies.

Howard began buying equipment in November 1993, starting with computers. These purchases, the contract stated, "shall automatically become the sole property of the [federal] government." A contractual amendment provided for up to $48,000 for lab equipment and up to $14,000 for office equipment.

Just after Christmas, the contract was modified, for the fifth time in five months, to add $39,000 to "curate and conserve the remains." A month later, Howard asked for and received another $34,000 for "additional services to curate and conserve the remains."

In February, six osteological technicians (OTs) were hired. Additional cost: $18,000. The total cost for the bio-anthropology work was already over $400,000, or about $1,000 per set of remains. An additional $97,000 was agreed to in May, "to continue preliminary cleaning and reconstruction of the human remains."

The true cost actually was higher: Howard requisitioned money as needed. GSA paid Howard and then added a large overhead for all monies sought for salaries and related costs. A $10,000 research bill thus could cost the U.S. Department of the Treasury as much as $14,000.

The GSA approved expenditures for the study and analysis of the skeletal remains. But it did *not* authorize expenditures for genetic analysis or for the chemical study of dental plaque to determine, if possible, where the individual was born and raised, as these studies had been deleted from the Research Design.

The work and the cost continued to advance. By August 1994, the first anniversary of the Howard-GSA agreement, the agreed-upon "firm fixed price" had jumped to $2,828,000, much of it for equipment. Blakey was now on the payroll for $250 per day.

The contract specified the "deliverables and final products" of the research venture and their deadlines: An initial draft report would be due by April 1, 1996. Then, "following GSA and Peer Review Panel comments" on it, "the final draft report shall be prepared and submitted" to GSA "no later than August 1 of that year."

GSA would have 60 days to comment on the work, which the contract stipulated as "the comprehensive report which is the major goal of this project." Howard University would then produce the report as a "finished product with no known flaws in terms of data, format, citations, editing [or] final illustrations."

Finally, Howard would prepare and deliver the Final Report to GSA, error-free and suitable for immediate distribution, by November 1, 1996.

With the project launched, the administrative controls were relaxed. The charter for the Federal Steering Committee, which had started out two years earlier as the Mayor's Advisory Committee, ran out and the group disbanded. The mandates of the peer-review panels of scientists also had ended, and they, too disbanded. Only one scientific link remained be-

SANKOFA?

tween GSA and Blakey:

Jerry Rose, the University of Arkansas anthropologist who had disinterred remains from the Cedar Grove Cemetery, was engaged by GSA as its consultant. He agreed to contact Blakey or visit him, when asked by the agency, to monitor the project's problems and progress. Rose said later that he had performed "precisely" the same sort of analyses at Cedar Grove (and elsewhere) that Blakey was embarked on at Howard, and so felt quite at home with this role. He said he explored specific situations, at GSA's request at Howard. But he had no mandate for overall oversight of the scientific work or time schedule. Neither did anyone else.

Blakey, meanwhile, recruited a staff. He drew heavily from his University of Massachusetts (UMass) classmates and colleagues. He chose as associate director his graduate school classmate, Lesley Rankin-Hill, who had obtained her doctoral degree in anthropology a few years after he did. Blakey would work closely with Rankin-Hill in the years ahead. But she never joined him on staff at Howard. Rather she remained at the University of Oklahoma, in Norman, where she eventually became an associate professor.

For laboratory director, his second in command, Blakey selected an anthropology graduate student, Mark E. Mack, who had earned his master's degree, *cum laude*, at UMass in 1990. Mack, in his early 30s, "expected" to obtain his doctoral degree from the University of Florida, at Gainesville, within the year (1993), according to his CV. Mack's area of expertise, in keeping with one of the UMass anthropology department's interests, was teeth and paleopathology: dental anatomy and the signs of illness, particularly anemia and malnutrition, that might be diagnosed by examining ancient disinterred teeth, jaws, and skulls.

Blakey and Mack appear to have worked well together. Mack was still listed as lab director and co-author with Blakey in scientific reports published more than a decade later. In 2007, he continued to be a Howard University employee, albeit he had yet to obtain the doctoral degree that he had said at the start he was close to earning. A relative newcomer to his professional field, in 1993, Mack's CV listed only three scientific publications.

Mack soon reported that 22 researchers were now at work on the African Burial Ground project, including, he said, 13 African Americans, 5 European Americans, 3 African[s], and one West Indian. In other words, most research team members were black — fulfilling Blakey's hard-won vision of who should perform these "sacred" studies. There was a startling exception, however: The only bone expert, or *osteologist* — the scientist who was assigned primary responsibility for handling, measuring, recording, and analyzing the ABG's mostly black human remains — was white: a southern white woman, born and raised in Alabama.

Her name is Mary Cassandra Hill. Colleagues and friends called her "Timmy," a childhood nickname derived from Dickens's character Tiny Tim in his "A Christmas Carol."

Hill, like Blakey, came from a medical family. Her father was a physician, her mother was a registered nurse, as were her two sisters and one brother-in-law. In her teens, Hill too, was headed for medicine. But she detoured.

Near her girlhood home in Alabama there are a number of massive — and awesome — man-made hills arrayed on a vast flood plain. They once were occupied by native Americans, and like similar constructions elsewhere in the central United States, they are referred to, unceremoniously, as *Indian Mounds*. They are tombs and platforms for ceremonial events,

SANKOFA?

and they were sacred to their builders and users.

Hill grew up in Moundville and visited Mound State Park frequently as a child. When she was 16, she obtained a summer job as a volunteer. A little later, she was offered a paid position as a guide and lab worker. She became captivated by the artifacts that she and others found at the mounds and the hints they provided of the lives of the long-dead American natives who built them. Hill's childhood interest became an avocation, and eventually, she decided to pursue anthropology, rather than medicine, as her career.

Hill is a slight, soft-faced woman, who dresses conservatively — with accents of crinoline — when she's not headed for a grave site or her lab, in which situations she wears a blouse and blue jeans. She is soft-spoken, affable, and shy — almost timid. But when engaged in scientific work, which is much of the time, she sounds competent and is unerringly determined. In a self-evaluation at Howard, she wrote somewhat later: "I essentially have two selves: one that is strictly business and one that is personable. I try to combine the two in a harmonious way, but when I'm at work, the business self is dominant." Both of Hill's "selves" were quickly engaged at Howard.

Hill was older than Mack, in her early forties. She, too, held a master's degree, and was a doctoral candidate, at UMass, Amherst. She also was a colleague and friend of Rankin-Hill, who had recommended her for the job. Unlike Mark Mack, however, Hill had more than two decades' experience in exhuming, preparing, and studying human skeletal remains and had two dozen scientific publications. Hill's appointment is startling in light of Blakey's earlier insistence that black scientists were uniquely qualified to study their forebears' remains. She would perform the most direct, hands-on work on these bones.

Blakey had wooed Hill strenuously. She clearly recalled, years later, having been summoned — urgently — from a classroom at the University of Alabama, at Tuscaloosa, where she was teaching physical anthropology, to take his phone call; he said he wanted her for the osteologist job. It was spring 1992. He asked for her input for the research design he was writing. A month later, Hill wrote Blakey, thanking him for his "confidence in my research skills" and for the "invitation to participate in your monumental project.... I humbly, and enthusiastically accept." She added, "As I understand our conversation, your request was for me to be a coordinator of pathology research, and to be a member of the pathology diagnostic team."

Hill went on to say that she would examine the skeletal remains for signs and symptoms of basic types of disease. She predicted that inflammatory, circulatory, and metabolic diseases might be "the three most important . . . for this population." Of particular interest would be analysis of [red-blood-cell-forming] disorders, which is to say *anemias.* The ABG individuals could have suffered anemia because of genetics — the inheritance of faulty red cells that is called *sickle-cell anemia* — or could have experienced anemia owing to nutritional (dietary) deficiencies. Anemia can be diagnosed even when the blood and soft tissues are long gone: In response to a deficiency in red cells, due to inadequate dietary iron, the blood cell-forming tissues in a youngster's body, principally tissues in the eye sockets and on the outside of the skull, overgrow abnormally. They create areas of porous, sponge-like bone that may remain at maturity. These anomalies are called (among other names) *porotic hyperostosis,* or porous overgrowth of bone. They tell a lot about an ill person's failure to thrive. They were and are one of osteologist Hill's particular research interests and the principal topic, later, of her doctoral dissertation.

In July 1992, in a "Dear Timmy" letter, Blakey said:

"What you sent was great[!]".

"This Design has been very favorably received so far," he added. "Meanwhile, our competitors, Metropolitan Forensic Anthropology Team/Lehman College have steadily lost credibility with The Community.... Despite all the deviousness on the New York end, rest assured that I will keep everything clean on ours. This is going to work!"

A year later, in July 1993, Blakey and Howard signed an employment contract with Hill. Her role was "to reconstruct, measure, and analyze [the ABG] human skeletal remains." She was to report to the lab director, Mack and she would have several OTs and a bevy of inexperienced osteological technician assistants (OTAs) to help her. Beyond her research, she was tasked with presenting the work to scientific colleagues and the public.

Hill had several projects to finish in Alabama. As a result, she was late moving to Washington. She arrived at Howard in March 1994 — when work on the remains had already started.

David Zimmerman

11

A Musket Ball and Beads

Lab director Mark Mack described the work in progress in the spring 1994 issue of *Update*: "The initial phase of the project," he wrote, "involves the cleaning, reconstruction, and data collection of the ancestral remains." Specifically:

After studying the New York field notes for a particular burial, the osteological technicians and assistants (OTs and OTAs) removed the skeletal material from the trays on which it had been stored and transported. By and large, each burial had yielded two "soil pedestals" of bones and soil, one containing the skull, the other the pelvis.

"The soil pedestals and other skeletal material are cleaned by dry brushing using artist's brushes, and gently removing the soil, using dental tools . . . . [An alcohol and water] solution is applied to . . . loosen the hardened soil matrix without damaging any . . . skeletal material." The removed soil was then shaken through a screen to look for shroud pins, beads, and other small artifacts.

SANKOFA?

Unmentioned in Mack's broad description is whether, and to what extent, he and his co-workers tried to protect the bones from contamination with their own DNA — breathed or otherwise shed from their bodies. Only a year earlier, scientific reviewers had said the best place to look for uncontaminated DNA was in the soil pedestals into which each cadaver's flesh and fluids had drained. It turned out later, according to Michael Blakey, that viable DNA could not be found in the pedestals — and eventually they were discarded. But DNA could be obtained by grinding up pieces of bones: Some 250 one- to two-inch segments were carefully sawed from long bones of the legs or arms and sealed away for later study.

Mack reported that the osteological workers were piecing the fragments together, using an adhesive, so that the bones "can be accurately measured and diagnosed." This is a key task, he said, in order to detect bone-altering illnesses. These reconstructions also facilitate measurements to determine each individual's sex, height, and age at death. The condition of the teeth provides complementary data. Together with the sex determination, Mack wrote, "we can reconstruct the demographic profile of the population.*

"It is important to know . . . whether people are dying at an early age, or if children are dying at a disproportionate rate; these determinations give us a picture of the proportion of men to women, the age structure of the community, and their health status."

As of April 10, 1994, the first 25 individuals had been processed. In this first sample, Mack cited these "interesting discoveries":

- Nine of the 25 were children, the majority under age six. In tough times, due to famine or illness, he noted, the

---

* But not the "race."

children suffer the most.
- Most of the adults had rotted dental cavities, growth disruptions in their dental enamel, and crowded, misaligned teeth.
- Thick, strong muscle attachments were found on some adults' long bones. These are marks of heavy labor.
- One young woman had died violently, a musket ball lodged in her rib cage. Mack said she had been shot in the back.

STILL, THE "MOST EXCITING discoveries" were not bones, but buried artifacts that provide "insights into the cultural practices of our ancestors." They include an unstrung line of blue glass beads (Burial #12) and teeth in two men's jaws that had

Cause of death. Musket ball (at end of pointer) came to rest under a rib, killing this young woman. Who did it is unknown.

been filed into sharp points and other shapes for ornamental purposes.*

Tooth filing, Mack noted, was a ceremonial ritual in parts of Africa.** "The filed teeth," he declared, "are proof that African cultural traditions weren't eradicated by the Middle Passage!"

Another such "proof," also from a person with filed teeth — and hence, presumably African-born — came from a middle-aged woman, Burial #340. She was wearing two strands of beads, one apparently a bracelet. The second, larger strand — seventy glass beads, one amber bead, and seven cowrie sea shells — girdled her hips. Besides their deep personal and symbolic value, such beads anchored African women's clothing, and, when shaken, were an erotic turn on.

Efforts to link these beads materially to Africa were less revealing. Most, if not all, were machine made, probably in Europe. One bead was of American Indian design. Project archeologist Cheryl J. LaRoche contacted a bead expert, Karlis Karklins of Ottawa (Ontario) Canada, the editor of the journal

---

* In public presentations, Blakey and his colleagues made much of the site's archeological richness and the thousands of artifacts that were found there. Less often said was that most of these artifacts came later, from the 19th century, when buildings over the burials were used as brothels. Jars and flasks were retrieved from their outhouses. While interesting in their own right, Blakey pointed out in the Research Design, these items say nothing about the black people buried there a century earlier. "Other than wood [coffin] remains, coffin nails, and shroud pins," he said, "there are very few artifacts are directly associated with the burials."

** Tooth filing is an African practice, Blakey maintained. Osteologist Cassandra Hill researched it and said no: Turks and other ethnic groups also filed their teeth, she said later, as did Hessian (German) mercenaries who fought on the British side during the American Revolutionary War. Thus, Hessians who died in the battles for New York City might well have been buried in the African Burial Ground (ABG). In the film *Sleepy Hollow* (1999), based on Washington Irving's story about the headless horseman, Christopher Walken, as a Hessian, displays ferociously filed, pointed teeth.

Beads. He was unsure of the beads' provenance. Asked later by phone if he could tell where the beads came from, he answered, "Not really! No!" Asked if they were likely made in Africa or North America, he said they are "more likely from Venice" or Amsterdam, trinkets commonly used by Europeans in the African trade. *

THROUGH BLAKEY'S MANY PRESS interviews in this period, and through Sherrill Wilson, and the Office of Public Education and Interpretation (OPEI) in Manhattan, the picture of the project's progress in the mid-90s was of slow, steady, forward motion. But when Cassandra Hill arrived in March, she was distressed to discover that Cobb lab was in disarray. Blakey and Mack were confused, she said later in an interview. They had the bones, but they didn't know how to proceed. They were not using the appropriate instruments to clean them. They didn't know what to do with them!

This was not surprising, Hill said. Blakey and Mack had little or no expertise in studying old bones. "Michael had very little experience below the jaw," she explained. "All you have to do is look at his curriculum vitae (CV) to see that!"

Hill insisted that the cleaning work performed before her arrival was inadequate — and would have to be redone. It was, she said later. But this did not endear her to her technicians, nor to her bosses, Blakey and Mack.

One of the only indications in his CV of Blakey's in-

---

* This was acknowledged, indirectly, when one apparently genuine African-made bead, found near Burial #434, was described in *Update* (Winter 2000) by archeologist Christopher R. DeCorse, PhD, an expert on the subject. "Other beads from the ABG are European beads that were widely traded in both Africa and the Americas during the 17th and 18th century," he wrote. "[They] could have been obtained by enslaved Africans after their arrival in New York."

SANKOFA?

terest in *old* bones was his claim to be a member of the Paleopathology Association: Members of this professional group study signs of illness (*pathology*) on ancient (*paleo*) bones. However, the paleopathologists' membership list for 1995 does not contain his name; neither does a supplement issued the following year. Hill, who says she went to all of the organization's annual meetings in North America, said she never saw Blakey at one of them.

Asked directly if Blakey did hands-on work in the laboratory, Hill replied: "No." She added that while Rankin-Hill "thought he was doing the [osteological] assessments, he was never there with me.

"Never!"

On August 12, 1994, Hill conducted her first osteological investigation of an ABG skeleton, a woman who appeared to have died in her early 20s. Hill recorded her findings by hand on several pages of personal notes, including a series of charts of the skull that she had drawn and copyrighted. She noted, among other pathology, that this Burial #1 had porous overgrowth of bone on the outside of her skull.

When she and her co-workers later turned to Burial #101, they made a discovery that Blakey then used to symbolize and represent the entire African Burial Ground (ABG) effort and its historical importance. The coffin held the bones of a man in his thirties, whose place of birth — African or American — has not been determined. It was his coffin lid, which was badly cracked, that made him stand out. Fifty-one large tacks (nails) stuck into it form a heart-shaped pattern. Within this figure, 136 smaller tacks appear to form different patterns, including the numerals "1769", very plausibly the year he died.

Later, Blakey explained how he made the identification: "I am [at] an African-American cultural event, and the pro-

gram had several symbols on the cover. And there it was! The symbol that I thought I had seen [on Burial #101's coffin lid].
... I took it to an art historian [Kwaku Ofori-Ansa] at Howard who specializes in this area. I tried my best not to appear excited! He, too, saw that it was some version of a ... Sankofa."

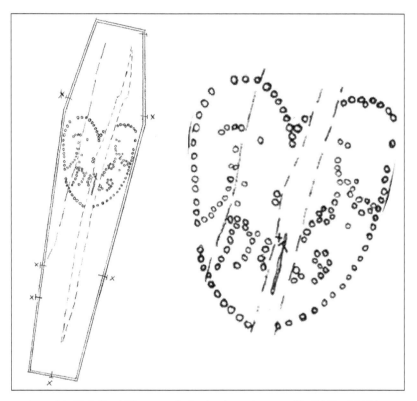

'Graphic Notation' Discovered. Design in tacks on coffin lid (Burial 101, a young man) was interpreted as an African mourning symbol called a Sankofa by lead scientist Blakey. Interior pattern indicates 1769.

THE SANKOFA HAS SINCE appeared, like a talisman, on all kinds of ABG materials, including letterheads and reports, and most dramatically as an ornament on the memorial at the cemetery's reinterment site. (See Frontispiece.)

SANKOFA?

A major part of Hill's work was overseeing and teaching the OTs and OTAs. She showed them how to handle the brittle remains, wash and brush away the soil pedestals to which they were attached, and match and mend bone fragments so that they could be carefully inspected and precisely measured. She and a photographer also took pictures of each bone from several angles. They made and stored some 60,000 images. Mark Mack set up thousands more pictures of teeth and jaw bones.

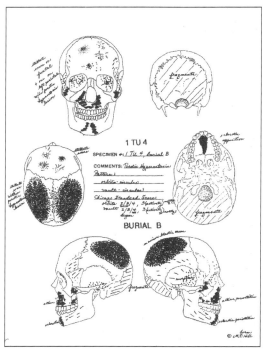

Recordation. Child's skull shows signs of porotic hyperostosis (stippling), in diagnosis by osteologist Mary Hill. But she was prevented from coding this data for *Final Report*. Images are from Hill's doctoral thesis at UMass Amherst.

HER BOSSES MIGHT HAVE been pleased with Hill for finally advancing the skeletal research that was their main rationale for moving the remains from New York. They might have been pleased. But they weren't.

Nevertheless, the work progressed: By autumn, more than 100 burials had been cleaned, reconstructed, and examined. A new report said that "stringent examination" of "the *ancestral* remains" had revealed that many skeletons had osteological abnormalities associated with manual labor. "Abnormal bone growth in the neck," for example, could be attributed to "the carrying of heavy burdens on the head" — which is the way many Africans and other members of non-mechanized cultures transport weighty loads. Similarly, "several cases of traumatic muscle pull involving the arms were found." The researchers continued to find signs of nutritional deprivation and developmental defects, as well as signs of severe infection. One woman "may have received a blow to the head."

Outside the scientific realm, the year 1994 saw a parade of distinguished visitors and ordinary people visiting the Cobb lab — some 200 had thus far signed the visitors' registry. They included a Nigerian cultural teacher, African leaders, and the director of a museum of slavery in the Bahama Islands, off Florida. Cobb lab, *Update* said, "has made a conscious effort to educate visitors about the ABG project, while simultaneously sensitizing them to the importance of African-American participation and [its] contribution to the study of biological anthropology."

In Manhattan, meanwhile, the OPEI was energetically conducting tours of the ABG site — now overshadowed by the federal office building being completed above it — and of the laboratories and repositories nearby where the non-human artifacts were being examined. Lectures, classes, and ceremonies illuminating black Americans' thus-far-unsung role in building New York City and the United States were featured. "Since our official opening in May of 1993," OPEI declared 18 months later, we have "conducted presentations and tours for more than 12,000 individuals!"

Under Sherrill Wilson's leadership, OPEI grew to be a vehicle of protest and hope — a major center of black activism in New York City. But, judged by *Update's* pages, OPEI made little or no effort to raise money independently to fund its various projects, which included, for example, a library reading room, films, site visits, and school programs. Rather, it largely depended on federal financing for these efforts, through John Milner Associates and the General Services Administration (GSA).

As OPEI phased in, in 1993–94, other sources of public information on the ABG project dried up. The GSA said little in public. The various committees and commissions, including the Federal Steering Committee, were gone. Blakey and, occasionally, Howard University spoke for the ABG project in Washington, and they facilitated laboratory visits and media coverage on the research in progress there. But OPEI was the main conduit of information. It quickly became, and long remained, the lens — and the filter — through which the media and the public perceived the ongoing ABG project.

David Zimmerman

12

Studying the Bones

A MAJOR SURPRISE, AT least for people outside the project, was the condition of the remains uncovered at the African Burial Ground (ABG). Photos of the opened graves, prior to disinterment, had been widely published in the press. Particularly Burials #355 and #356, a mother, who apparently died at childbirth and her stillborn or newborn baby. Also Burial #25, the woman with a musket ball in her chest. These skeletons were relatively intact, laid out much as they had been buried 200 or so years ago, which suggested to readers and viewers that many, perhaps most, of the recovered remains were similarly complete. But this was not so. Very few sets of remains were essentially intact. For many, all that was left were a scattering of bone fragments or, perhaps, a handful of teeth. Thus, there was less human material to work with than may have been expected.

This "generally poor level of skeletal preservations," Mack

SANKOFA?

and Hill wrote in a report, resulted from "natural deteriorations ..., exposure to ground water seepage, pressure damage from the 16 to 30 feet of landfill [that covered the graves for more than a century], building construction disturbances... [in] the 19th and 20th centuries, and excavation and curation damage in New York.

"As a result of these destructive forces," they added, "the teeth, which are the hardest structures in the skeleton, are often the only remains found, and therefore are very important for our research .... teeth often provide the only testimony to the [decedents'] life conditions."

One of the clearest tales that teeth tell is their owner's age at death. They reveal it with errors of less than one year, according to Mack. Based largely on the dental data, he and Hill say, "one half" of the "approximately 400 skeletal remains" were infants under age two.

This was information that later analysts and interpreters of the ABG data would have to struggle with. Certainly, early death was common. But did these findings reveal anything significant about the black population's age profile in 18th-century New York? That was — and is — a difficult question. To cite just one problem: Was the age profile of the disinterred 400 sets of remains representative of Manhattan's black population as a whole? Or even of the ABG as a whole? It might not be.

At another historical Black cemetery that was just then — in the mid-90s — being excavated, in Dallas, Texas, the burials were *not* randomly distributed by age. At this Freedman's Cemetery, where some 1,100 human remains were excavated to make way for a road, older people were buried close to the church, while children under age five were elsewhere.* Con-

---

* Young people customarily were buried in the "child's garden" at the side or back of cemeteries, according to Hill.

ceivably, to spare pallbearers' backs, older, heavier bodies were buried close to the church door, while younger, lighter ones were easily carried farther afield.

Were the ABG interments randomly distributed or, rather, grouped by age? Mack and Hill had no way of knowing. The ABG archeologists later would say that men, women, and children were distributed more or less evenly across the excavated area. But "there are a few places where numerous children's and infants' graves seem to cluster, . . . Sex distribution [also] is skewed, with a preponderance of men in the northern part of the cemetery."

Far more certain were the data that could be discerned from the individual remains themselves, particularly from the teeth. They "provide an indelible record of a child's health status," Mack says. Particularly revealing, he and Hill add, are flaws in the tooth enamel. "Our available . . . data show that 50% of the children . . . experienced stressful health episodes, as indicated by enamel defects."

Specifically, they noted defects in tooth formation that are manifest as rings or pits of poorly formed enamel (*enamel hypoplasia*), as well as disruptions in the uptake of calcium and other minerals into the teeth that are manifest as brown or other discolorations (*enamel hypocalcification*). Many such defects reflect poor nutrition or other health problems, they say. Burial #7, a four-year old, had poorly formed enamel on the canine teeth, suggestive according to Mack and Hill, of poor maternal health and breast-feeding problems. Burial #43, another four-year old, had "severe teardrop-shaped pits [in the teeth] which represent a prolonged stressful episode" before death.

Besides the enamel defects, one-third of the children had cavities. Poor diet was "a major causative factor," the re-

SANKOFA?

searchers say. "Instead of benefiting from a varied, nutritious diet, enslaved Africans were relegated to a diet . . . of nutritiously poor foods."

Bones, too, were compromised. Mack and Hill point out that tooth development is largely under genetic control, and so proceeds apace. Bone development, however, is far more dependent on environmental factors, particularly diet and illness, so that the difference between one's dental age and one's bone age at death is a further measure of hardship. "Most of the younger children show skeletal growth retardation relative to dental development," they write. Many show a one-year lag between dental and skeletal ages. Some children under five have two- to two-and-one-half year lags in skeletal development. From these and their other findings, the scientists conclude:

> The dental and skeletal indicators reveal information that is not written in the history books concerning our ancestors: Namely, the effects that forced enslavement, unpredictable conditions and poor socioeconomic status had on the youngest members of this population, as evidenced by pathological observations. By no means were enslaved African children shielded from these dreadful conditions. Quite simply, they were most severely affected because they did not survive into adulthood. The skeletal remains of these unfortunate children are a lasting testament to our past.

In this conclusion, however, Mack and Hill go beyond the scientific data they have so carefully collected. They offer no evidence that the children and their parents were in any worse shape than the white settlers and colonists among whom they lived. Childhood illness and death were common in 18th century North American communities. Mack and Hill may be 100% correct in what they write. But, lacking comparative data from white cemeteries and Native American burials of that era, the assertion that "enslavement" caused the physical damage in these "unfortunate" children's skeletal remains isn't science; it's simply conjecture.*

What can, of course, be said with certainty is that, without slavery, and the roles of the black Africans and white traders and colonists who profited from it, the ABG people would not have been in Manhattan in the first place. Would their health and well-being have been better had they remained in Africa?

Blakey, meanwhile, had serious administrative concerns in 1994. For one thing, his deadline for completing the research and writing the first draft of his report loomed large ahead: April 1996. Also, despite constant infusions of federal money through the General Services Administration (GSA), he kept scraping bottom. Manpower was short, some vacancies were not filled, and some needs were not met, he recalled later. Blakey also said later that he did not take payment for many days that he worked, in order to preserve funds for absolutely necessary scientific expenditures.

These shortcomings may seem strange, given the rate at which Blakey requested — and received — money from the feds: In January 1994 the GSA/Howard contract had been modified six times, and the amount authorized had risen to $390,000. One year later, in the contract's 14th modification, this amount had risen nearly 10-fold, to $3,612,000. It would continue to rise. But the cost already was approaching $10,000 for each set of human remains.

One thing Blakey held out money for was a genetic study of the remains — which GSA declined to pay for. Fortunately, one of the four resident Howard University ABG researchers

---

* Later comparisons with remains from nearby Trinity Church Cemetery confirmed that the blacks' health and longevity were poorer than whites'.

SANKOFA?

was well-qualified for this effort: Biochemist Matthew George, PhD, began in 1995 to test "the feasibility of isolating 'ancient' DNA from the remains," which will allow us to determine [their] possible sites of origin. A "small portion of a GSA grant was used" for this purpose.

It proved to be difficult and frustrating work.

Meanwhile, correspondence between Blakey and GSA officials shows that all was not well between them. In a Progress Report on March 3, 1995, Blakey complains, for example, that "x-rays and other 'specialized' studies have not been done due to delays and restrictions in GSA and delays in allocation of space by Howard . . . ." Blakey passed the buck, telling GSA that "the x-ray machine has not been in use due to an inadequacy of space which Howard . . . officials have repeatedly indicated they would resolve." But hadn't. "Both Howard and GSA appear to bear responsibility for [the] delays," he said.

Update's summer 1995 issue reports that Blakey had been awarded an honorary degree by York College in Queens, New York; York is a part of the City University of New York, against which Blakey had competed — and won — the right to study the ABG remains. The honorary degree cites "Dr. Blakey's outstanding scholarship in biological anthropology, bio-archaeology, paleo-pathology and ethno-history" and "the positive influence this research has had on students considering . . . anthropology . . . .

"Dr. Blakey serves as a good role model," the citation says.

A major goal in 1995 was to begin presenting Cobb lab's research findings to colleagues in the scientific community. The venue Blakey chose was the annual meeting of the American Association of Physical Anthropologists (AAPA), to be held in Oakland, California, in March. The Cobb group would present several oral reports of their findings. One, written by

Hill, was titled "Women, Endurance, Enslavement: Exceeding the Physiological Limits."

Hill drafted it. Embedded in an historical narrative, she and her co-authors describe skeletal evidence of severe overwork:

> In this population, muscle attachment hypertrophy [excess growth] is common among all bones of the arms and legs. What is perhaps most striking . . . is the extraordinary incidence of pathology of the muscle and ligament attachments. Seventy-five percent of males and sixty-five percent of females exhibit these pathologies . . . [that] result from strain upon muscles and ligaments sufficient to tear bone away from their attachments. The[y] . . . represent . . . work and load-bearing stresses at the margins of human biological capacity. The skeletal distribution of these lesions suggests behavior related to the biomechanics of lifting and carrying excessive loads.

Talking to Blakey about the paper while she was writing it was difficult, Hill said later — particularly when she tried to discuss the ethnicity and origins of the burials. She noted, based on data from the Library of Congress and New York Public Library, that people of various groups were buried together according to skin color. Dark people from India, brought by the Dutch, were buried in the cemetery, as well as Africans, she noted. So, too, under the English, were white non-Anglicans and people who had not been wed in the Anglican church. Prisoners also were buried there, including executed convicts, as were Hessian soldiers who fought alongside the English during the Revolution. Given the "rich history" of this cemetery, Hill said, Blakey's "narrowness of focus" on it, as a *black* cemetery, "bothered me."

"I talked to Michael about some [of the burials] I thought

were Native-Americans, not African-Americans. He just wouldn't hear it. The more I tried to talk to him, the more rigid he became!"

Blakey told GSA, contradicting Hill: "We have no indication that any [one] . . . from the burial ground [is] Native American."

Aggravating Hill, Blakey made plane reservations, through Howard, for himself and Mark Mack to fly to Oakland for the AAPA conference at which Hill was to be a presenter. Hill was not included in the reservations. When she protested, Blakey told her to make her own reservations with her personal credit card. She did. Subsequently, she said, the ABG project and Howard University both refused to reimburse her.

Much later, it turned out, from GSA documents released under the Freedom of Information Act (FOIA), that Howard University's contract with GSA did not provide funding for staff members' travel. Blakey protested. He tried — and apparently failed — to get reimbursement for Hill.

Meanwhile, Matthew George now had initial results from his genetics study. He had successfully extracted DNA from the bones of several ABG individuals and had begun to determine their genetic sequences. (See below, Chapter 18, for an explanation of these experiments.)

George reported one other noteworthy fact: A second Howard University geneticist, who was not part of the ABG project, had run separate, parallel tests on some of the bone samples. His early findings were largely commensurate with George's — which supported the notion that the scientific methods to isolate and analyze the long-buried bones' DNA were valid.

13

Loggerheads

Very quickly — within a few months — Michael Blakey and Mark Mack had found themselves at odds with Mary Cassandra Hill. They didn't like her behavior.

"I was going outside my employment boundaries," Hill quotes Blakey as saying.

In Blakey and Mack's view, she was bossy. She was trying to take over. She didn't know her place in the project! Hill denied this, saying she was within her written contractual bounds. She did not want to take over — wanted, rather, for all of them to complete the project successfully. Blakey became furious at her.

Hill began to record their interactions, using little slips of blue paper that she kept in her desk diary. By Christmastime 1994, she had processed two-thirds of the 400 sets of skeletal remains. "On December 22nd, at 5:30 p.m.," she wrote on a blue-paper note:

SANKOFA?

Michael Blakey called me into his office. He handed me a cup of heavily spiked eggnog and said, "Here. Drink this. You're going to need it for what I'm about to say to you." He then delivered the worst verbal assault I've ever received. He accused me of trying to take over his project. He said I had insinuated myself into every possible aspect of the project. He told me that I didn't know my place. He said that people were constantly complaining about me. Each time I tried to defend myself his tone and behavioral gestures became more aggressive and animated. He used very foul language.... He accused me of telling people ... about the problems at the project.

When I denied all of this, he yelled at me. He said he didn't "give a flying fuck" what I had to say in my defense. He told me that I was "like a disease!! ... You're Pathological!" He said that I had a history of emotional and psychological problems. I denied this and asked for substantiation. He said he didn't have to. He said that I was "very poorly socialized," and that I had no friends. I told him that that wasn't true, and that I felt that several people on the project were my friends. He said that that was not true. "Don't you understand?! WE DON'T LIKE YOU!"

This verbal abuse continued for 3½ hours. I was crying. Finally, he stopped and said, "Now. Lets talk about your teaching osteology in the spring." I was shocked. I said I would not teach his osteology course and that I wanted to leave. He insisted on walking me to the door since it was late.

I cried all night. The next day, in the lab, everyone wanted to know what had happened, since I looked so bad. Several people tried to console me. Michael was there and was still acting in a very angry and aggressive manner. He told me I needed to go home for Christmas and "give serious consideration to all that I've said" [emphases in the original].

David Zimmerman

After Christmas, Blakey continued in the same vein:

> On January 11, 1995, [he] called me into his office at 10 a.m. He said that during the holidays he had had the opportunity to review the evaluations that everyone had filled out (actually several people refused to fill them out) and that, by far, mine were the worst of all. He said there were no positive responses about me. I told him I didn't believe him and asked to see them. He refused. He said he had written down all of the responses that he said were the most important and the most critical. He read these to me. I tried to deny them and to defend myself, and he said that he wasn't interested in anything I had to say.
>
> He handed me a yellow legal pad and instructed me to take notes so there wouldn't be any misunderstanding what he said. He said I was to be punished. He said he wasn't going to fire me, although he felt he needed to. He said he would [not] fire me because he had known me a long time. He listed several items to which I was to adhere. He accused me of trying to eavesdrop on all of his business, and said that I was to stay out of the office. The copy machine, pencil sharpener, & stapler are in there, but I have to knock on the door and ask if I may have access to them. I told him I wasn't trying to eavesdrop. He said he didn't want to discuss it. He said I'm trying to take away his project.

Blakey then handed Hill a directive that said:

- Cease "encroachment" on lab director's and others' work.
- Stop making derogatory, demeaning, or misleading comments toward and about project staff.
- Follow procedures and policies as directed by lab director Mack.

SANKOFA?

- *Stop encroaching on others' responsibilities.*
- *Stick to your task of recording data from the remains — and don't slow down.*
- *Leave the lab by 6 p.m., unless otherwise advised.*

Blakey docked Hill a month's pay, illegally, she said later. He told her that she was now on probation.

One unsaid rift may have stood between them: Hill was working on her doctoral degree at UMass Amherst, as Blakey had. But her earlier studies, for her master's degree, had been at the University of Tennessee in Knoxville. The focus of that school's biological anthropology faculty was anathema to Blakey:

The University of Tennessee conducted what may be the foremost — and certainly is the best known — program on the legal and medical issues involving human bodies, the discipline called *forensic anthropology*. They train crime scene investigators (CSIs), including pathologists, medical examiners, and agents from the Federal Bureau of Investigation and other law enforcement agencies who examine crime victims and other suspicious deaths. Later, these trainees often testify in murder trials.

An ongoing and well-publicized venture of this anthropology department is its so-called "body farm," where cadavers are simply laid out on the ground so that researchers can study the decay process through which the human body unto dust returneth.

The University of Tennessee course of study for graduate students is quite rigorous, Hill said. "My training in osteology, from forensic anthropologist Dr. William Bass, PhD, was *extremely* strict," she recounts; "for instance, we had to identify fragments of bones inside a box . . . by touch alone, as a weekly bone quiz was part of our final grade" [emphases in the original].

In this forensic approach, the traditional information sought from an unidentified body was sex, age, stature, and race — and in the last, the "race" of a victim and of suspects typifies an approach to law enforcement that, as Blakey rightfully said, tends to identify and victimize black suspects and defendants. Blakey's "biocultural" approach to the African Burial Ground (ABG) remains was very much like his African-American reaction to forensic specialists' white, racist treatment of black bodies. He may have come to see Hill as the embodiment of this despised approach, which then became a cause of friction between them.

As Blakey wrote later, "forensics work relies on the objectified categories of biological race identification, without relying upon (or constructing) social, cultural, and historical information that is at the core of the biocultural approach" that he preferred.

The enmity seems to have arisen, in part at least, because Hill is white. And a woman.

Hill later denied Blakey's charge that she was trying to steal the project. Her interest, she said, was its overall success. Her interest did, however, extend beyond analysis of the bones. On her own time, she researched blacks' living conditions, as far back as the Dutch founding of New Amsterdam in the 1620s.

She had found that New York blacks had a "much richer" history than had been generally acknowledged. Among other things, she discovered that after 20 years of slave labor for the Dutch West Indian Company, the families of the first 14 slaves (11 men, 3 women) were granted conditional "freedom" and "were at liberty to hire themselves out when not working for the company . . . they could travel freely throughout the area . . . [and] were 'free' to supply their own subsistence on their small homesteads."

Blakey, a co-author on the report that Hill was preparing for the American Association of Physical Anthropology (AAPA) meeting, did not like these nuances on the enslaved Africans' living conditions. He told Hill to delete this material. She did.

While still shy on data, the oral reports that Blakey, Mack, and Hill presented to AAPA, which were abstracted for its journal, were among the ABG scientists' few efforts to fulfill their agreement to keep their scientific colleagues abreast of their discoveries.

By mid-March 1995, Hill had turned to Burial #25, the woman with a musket ball in her ribs. The lead projectile was flattened on one side, suggesting that it had struck a rib, causing a mortal wound. Hill determined that the shot came from behind, or, alternatively, from under the woman's armpit.

Hill said Blakey was trying to create a scenario in which the woman was shot trying to escape. She demurred, saying the shooting might have been a domestic affair.

She said the woman also had been beaten in the face, leaving what is called a *Le Fort fracture*. "They had never heard of it before," she said later.

The ultimate source of information on wounds from bullets and other missiles was the Armed Forces Institute of Pathology (AFIP), a Department of Defense agency in Washington, that is now defunct. Having examined the skeletal remains of a number of people killed by gun fire, Hill decided to check out her findings on Burial #25 with an AFIP expert whom she knew. Blakey was not pleased, she noted on a blue slip in mid-March:

"[I] went to AFIP today. I saw [my friend]. He helped me with references for my presentation . . . Michael yelled at me . . . . He said I'm not following his instructions. 'You will do *exactly* as I say. DO YOU UNDERSTAND?!' He said he wants *Mark* to be the one to develop a relationship with AFIP. I said O.K., but I'm not going to ignore my own long-standing friendship with [my colleague] in order for Mark to do so" [emphases in the original].

Interestingly, Burial #25 turned out to be the *only* one of the 400 that could be shown from the remains to have died from interpersonal violence — hardly the finding one might expect, given the violence and inhumanity Blakey insisted blacks had been subjected to.

At one point, Hill recalled later, Blakey asked her to "reassess" her findings and look for necks that had been fractured in executions by hanging. She refused. She said she knew quite well from her earlier work what a hangman's fracture looked like, and had not seen any from the ABG. She told him that she stood by her findings and that if they now discovered other presumed cases of deadly force, they could be ridiculed later if their evidence proved unpersuasive.

Hill is not the only scientist who appears to have been pushed to find evidence of mistreatment that she herself had not discerned. Anthropologist Leslie E. Eisenberg, said later that, while working with Blakey at the ABG site in 1991-92, he prompted her to interpret infectious marks on denizens' ankle bones as evidence that they had been shackled in life. She declined to do so.

Blakey was looking for the remains of blacks who had been legally lynched near the ABG in two dramatic incidents: trials following a slave insurrection in 1712 and an alleged slave conspiracy in 1741. In 1712, 20 black men were hanged and three were burned at the stake. In 1741, 17 black people were hanged and 13 were burned at the stake. None of the remains has been found.

Relations in the lab did not improve. By June 1st, Hill had processed most of the remains. She wrote:

> Today was my birthday. I put on a new dress and went into work as usual.
>
> I had hardly sat down before Michael came in and started yelling at me about burial #25. He wanted to know why it was still on the table. I told him that I was trying to do a really complete evaluation on her since she's special. He countered that they're all special. I said I knew that, but that she's particularly special since he (Michael) makes a point of mentioning her on all special occasions — so does Mark — so does everyone who gives tours. He was clearly angry about something else, but used this to vent his anger. He went away and came back yelling — five times in all. [Other people] were present. They were stunned. Michael didn't care that they were there. I was reduced to tears.

He, meanwhile, had gone through official channels at Howard to have Hill declared unfit. He told her that he would fire her if she didn't go to a psychiatrist, she said, and he sent her for counseling. She agreed to go. The counselor, Silas L. Parrish, MSW, who was program chief of the DC (District of Columbia) Employee Consultation and Consulting Service, a government agency, wrote back to Howard:

> Ms. Hill was seen for registration at this facility on April 14, 1995 in response to your referral. [She] was seen by this writer, and I found her to be relaxed, cooperative, and open. She appeared open in her presentation and responded thoughtfully to questions.

> *I could detect no significant problems in Ms. Hill that would preclude her ability to function satisfactorily on the job. However, in order to have a more in-depth evaluation, I recommended . . . an evaluation by our staff psychologist. Ms. Hill accepted . . . .*

Following her evaluation interview, the psychologist reported:

> *There does not appear to be any significant psychological/emotional problems. Ms. Hill appears to be an over-achiever, always driving herself, working all hours, including Sundays. As a dedicated scientist she seems to expect those with whom she works (and supervises) to be as dedicated as she. It was recommended that Ms. Hill participate in a period of individual counseling (15 sessions) to help learn to cope with a situation which she finds stressful and threatening. Ms. Hill indicates that she likes the project and wants to continue. The counseling may help see her through the balance of the time. Ms. Hill was in acceptance of the recommendation . . . .*

In August, the therapist whom Hill had seen, Lori Heberley, MA, wrote her: "Just a quick note to tell you how much I enjoyed our relationship over the past few months. I always looked forward to our appointments and your company. I wanted to tell you this yesterday but I was afraid that it was 'unprofessional'. However, getting a lot out of being 'unprofessional' seemed to be yesterday's theme, so I thought I'd write you this letter! Best of luck to you in all that you do! [I] will think of you fondly and often."

Blakey continued to complain. What he did not do, Hill asserts, was roll up his sleeves and join her and the technicians at the examination tables.

SANKOFA?

GSA consultant Jerry Rose, in an interview, concurred with Hill:

Blakey did not do hands-on work.

Lab director Mack also was limited, Hill said. "He only did teeth."

She explained that by that point in her career she had exhumed, stabilized, and studied hundreds of human remains — and she knew exactly how to proceed. And she did.

"Michael had no experience in how to do this kind of job, and neither did Mark Mack." She added, "I had my own equipment. I'd done this kind of work for years!"

Drawing soothes Hill. *New York Times* writer Brent Staples spoke to both Hill and Blakey in Washington, and handed Hill this cartoon. But he did not report the project's problems in his newspaper. Neither did anyone else.

14

RACISM

NOT A WORD WAS published about the Cobb Laboratory's internal troubles. Few people on the outside knew about the antagonism between Michael Blakey and Mary Cassandra Hill. Or, if they knew, they weren't saying so in public or in print.

There was one exception in terms of *knowing* what went on: Brent Staples, a writer for the *New York Times*, who is black, visited the lab and spoke with both Blakey and Hill. "He was always really nice to me; I like him a lot," Hill recalled later.

"Once when I was on the podium [to speak] in Washington, he came from his seat in the audience, and handed me [a] note, and told me to smile, and be confident." Staples wrote in the *Times* about the ABG Project's prospects but *not* about its problems.

Blakey was having trouble from several quarters. Early in 1996, for example, the General Services Administration (GSA) had asked him why his expense sheet listed $17,750 for telephone calls. Later a contracts official wrote to complain that his first draft was overdue and that a long overdue progress report had not been sent — despite three agency reminders. "Please provide these documents without delay," the official said. "Should you have any questions . . . we are, as always, willing to assist Howard in expediting this process."

The official also told Blakey that his cash was short because GSA had approved payments of over $5 million — but as of June 1996, Howard had only billed the agency for $550,000.

"Howard for years couldn't figure out how to send us an invoice," recalled Peter Sneed, the GSA official who dealt with these matters. At one point, he added, "22 months went by without them sending us [one]." Sneed said: "I have not a clue why they couldn't [send out a proper invoice]."

Blakey thus seems to have had some trouble with his home institution. To cover his deadline failures, he asked for contract extensions of the due dates. But sometimes, Howard failed to follow up. At one point, GSA wrote: "Your latest Progress Report to us, dated November 31 [sic] 1996, stated that this request for extension would be sent in a few days. More than two months have passed, and we have yet to receive [it]. I cannot over-emphasize how critical it is that this request . . . be submitted immediately, in writing."

Blakey, meanwhile, pridefully elaborated on his scientific rationale — his purpose — in a key paper co-authored with African Burial Ground (ABG) research associate Cheryl J. LaRoche, PhD; it was published in 1997 in the journal *Historical Archaeology*. The title and topic: "Seizing Intellectual Power," in order to define and control the burial ground's meaning.

The site's "significance," Blakey and LaRoche declare, should be understood in relation to blacks' "vindicationist" efforts. *Vindicationism*, they write, was originally set forth in 1929 by Arthur A. Schomburg, the black scholar for whom

New York City's Schomburg Center for Research in Black Culture is named. The approach, as described in Blakey's Research Design, is intended to reflect black Americans' "critical intellectual, education[al], and political concerns."*

Vindicationist studies aim to correct the distortions of earlier research that were used by whites "either to systematically victimize or, alternatively, to ignore" black people, Blakey declared later. "By omission, northern slavery and racism were denied." Vindicationist science, as Blakey described it, thus was — and is — grounded in racial conflict and racial politics: It starts with a pro-black, anti-white point of view. In other words, it is racist — *black* racist.

Looking back to his conflict with the Municipal Forensic Anthropology Team (MFAT), Blakey and LaRoche said:

> *When vindicationist motivations were explained as part of the site's significance for the African-American community, Euro-Americans . . . expressed fears and objections, characterizing the approach as ethnocentric bias. Yet the vindicationist tradition was posed as a corrective for persistent Eurocentric bias and misrepresentation, and as a search for truth and accuracy.*

Anthropologist Leslie Eisenberg, formerly of MFAT, agrees with Blakey that the people in power write the history. But, she adds, this was all the more reason why the analysis of the remains should be conducted as dispassionately as possible, so that the skeletons could tell their own stories, ruling out bias from both white and black researchers.

"The bones have not been biased in this way" by observers,

---

* Elsewhere, Blakey attributes vindicationism to an earlier activist-scholar, Carter G. Woodson, in 1915.

SANKOFA?

she said. "They are an objective source of information, if they are treated objectively!"

Blakey, nevertheless, had scored points against MFAT because it lacked blacks. So, he declared, MFAT was *racist* — and it had no comeback beyond saying that its shortage of black members was inadvertent.*

In 1994, one of Blakey's associate directors for the project provided a new perspective: Archeologist Warren T. D. Barbour, PhD, of State University of New York at Buffalo, a black man, explained in a National Park Service publication, *Federal Archaeology Report*, that there were then fewer than five African Americans with doctoral degrees in his specialty (of whom he was the first).

Why so few? *Not* because of racism.

"Thinking back over the three decades of my career," Barbour explained, "I have come to the conclusion that there is no conspiracy behind these small figures. Rather they arise from the history of archeology as a discipline, combined with the aspirations of my ancestors."

His own career experience, as a middle-class African American, may have been typical, he added. "My family was aghast when I decided on archaeology as a career. I was expected to enter one of the traditional middle-class Black professions:

---

* Is anthropology "racist," as Blakey charged? At the start it certainly was. But Blakey's close colleague Lesley Rankin-Hill, PhD, says pioneering anthropologist Franz Boas, PhD, "was the first . . . to seriously advocate a research focus on Afro-Americans during the early 20th century [in 1906 and 1909]. He proposed the creation of an 'Africa Institute' of anthropological research focusing on the accomplishments of Africans, anatomical studies [of them], and statistical analyses.

"This was . . . to eradicate racism by showing whites the positive attributes of Afro-Americans' African ancestors, and [by] disproving the biological, psychological, and moral inferiority myths concerning them, and to improve the Afro-Americans' self-images and conditions by making them less 'despondent', more hopeful, more proud of their heritage, and ambitious to change their conditions."

lawyer, doctor, minister, or undertaker. Middle-class families like mine," he added, "when envisioning something better for their children, embraced the American values of status and money, along with the professions that would deliver them."

The same low numbers and reasons appear to have prevailed in Blakey's field, *biological anthropology*.

In their "vindicationist" report, Blakey and LaRoche spell out why they rejected the statistical method for determining an individual's ethnicity, or "race," from measurements of the bones, as MFAT and others had proposed. One such method, developed by MFAT anthropologists James Taylor and Spencer Turkel, was based on detailed measurements of just a single bone, the femur (thigh bone), and had been reported to be 94 to 97% accurate in identifying the ancestry of skeletons of known lineage.

In their retort, Blakey and LaRoche say that "Euro-Americans'" anthropological studies of black people have historically been used to make the assessment that whites are superior and blacks inferior. "For African-Americans today," they say, "'racing' has been associated with arguments in support of black inferiority, social and biological distance, and stereotypical images . . . ."

This method, they charge, assumes the existence of "a real racial biological type." Among its traits: jutting jaws, large teeth — the kinds of traits that were being described to the public.

Reliance on measurements of the hip and thigh bones (innominates and femurs) is demeaning, and not validated as science, they add. Most important, these analyses are "dissociated" from the particular cultures and history of the ABG's population. In sum, LaRoche and Blakey say: the Descendant Community understood "the parallels between the mishandling of the bones and the racial reality of their lives. If race follows the African descendant population beyond the grave, then racism, "by definition, follows as well."

For Blakey, the remedy — *his* approach — holds "greater relevance" for black people, and what is more, it exposes the "biases" of "mainstream" or "Eurocentric" studies.* This is why, he and LaRoche add, opponents had denounced their approach as ethnocentric anthropology.

The "vindicationist" approach, they add, serves as "a corrective for persistent Eurocentric bias and misrepresentation, and as a search for truth and accuracy.... Seizing intellectual control has meant that the criteria for competency have been expanded to include an affinity for African-American culture...."

From this perspective, they insist that the only way to characterize the remains would be to determine, simply by looking at them, if they had been African-American people — a nonscientific method. Blakey suggested elsewhere that he wanted to study them in terms of their specific cultures and histories.

Meanwhile, Blakey's personal view on "race" and on white people was summed up neatly in the name he and his wife chose for their son: *Tariq*. That's the name, he explained later in *Current Biography Yearbook*, of the African warrior who conquered Spain in 711 A.D.

---

* Blakey's insistence on Afro-American or African-American designations for black people is, of course, a rhetorical device that many reject. As the black social critic Harold Cruse had written earlier: "The term Afro-American suggests and designates the Negro's African background; but it is also another way of expressing non-identification with the American Negro *qua* American Negro and his social status in America.... There is now a faction in Harlem who want to dispense with Afro-American as too 'moderate'— in favor of African-American. Then there is the ultimate integrationist trend that prefers simply American, without hyphenated qualifications (not to speak of those extreme nationalists who maintain that they have never ceased to be Africans despite centuries of separation from that continent.) The problem with these various schools of semantic nationalism is that they merely create more conflict, while solving absolutely nothing concrete in the way of political, economic, or cultural advances for the Negro in America. Their word battles are the results of accumulated frustrations and social blind alleys...."

15

WAR

MICHAEL BLAKEY WAS NOT assuaged by Cassandra Hill's clean bill of psychological health from Howard University. What had been a battle became a war.

By May 1, 1996, Hill had processed 373 of the 427 sets of remains. All that were left were 27 that had been quarantined because they were infected with dangerous microorganisms, and 27 others that Hill had saved for last because they appeared to be particularly interesting, due to severe infections or injuries during their lives. Nevertheless, Hill still felt unrelenting pressure from Blakey and Mark Mack. She decided to file a grievance against Blakey. She wrote to him requesting a grievance meeting, with Howard officials participating, to address her "concerns" and "hopefully [find] a successful resolution regarding":

- The work environment, in re interpersonal relations.

Sankofa?

Together. Michael Blakey towers over Cassandra Hill as they pose for group photo in Howard University lab.

- Work in progress.
- Her professional work outside the scope of the project that may be "indirectly/directly" associated with it.
- Her concern for recognition as a "major contributor" to the project.

In the last regard, she noted that he, Blakey, and Mack were moving ahead on a research paper "for which I am the senior author ... without me." She thus broached two volatile scientific issues — of intellectual property rights and authorship — that would heat up quickly and then smolder .... They still smoldered, a decade later.

BLAKEY WAS NOT PLEASED. Hill noted in her 1996 daybook, on May 14th: "Michael came into the lab and harassed me, grabbed up [a] folder & read my notes, [and] admonished me for commenting on mandibular [jaw bone] pathology." She added, perhaps disingenuously: "He was very angry about something."

It seems likely that that *something* was her grievance.

The meeting on it had been convened a few days earlier by Florence Bonner, PhD, the head of Howard's Sociology and Anthropology Department. Human Resources official Margaret V. Sullivan, also was present. The school's meeting lasted 4½ hours. It was "very painful," Hill wrote in a daybook note on May 16. "Michael denied everything ... as if it was all my fault."

Bonner mediated the session and summarized Hill's complaints in a memo:

- Blakey had harassed and intimidated her.
- He had shown unprofessional conduct by berating her in public.

- He discriminated based on sex by treating Mack, a man, differently, speaking to him in private when there was a problem.
- He refused to let Hill meet with consultants for the African Burial Ground Project (ABG) who were visiting the lab.
- He threatened to publish work of hers without her permission. . . .

Blakey responded to each complaint and was supported by Mack. Bonner wrote:

"Dr. Blakey said he was not aware that he had indeed berated or harassed or intimidated Ms. Hill . . . . [We] were able to agree that he had approached her in public and expressed displeasure with her work . . . . [H]e might have been 'firm' . . . with Ms. Hill . . . because she 'continues to try to do everything,' and as a result, she was not completing her specific tasks."

Hill's complaints were seconded by Human Resources counselor Sullivan. She told Blakey that "at least four other females had made exactly the same complaints about him . . ." She added that "after hearing the description of [the lab's] task assignments and lines of authority, she was surprised that [any] work was getting done."

The resolution was that Blakey would "speak with Ms. Hill in private if he had a problem(s) to resolve" and "he would take care with how he in fact spoke to her."

A couple of weeks later, Hill noted, "I've spoken to Mark Mack several times since 'the meeting.' He has behaved in an extremely rude & immature manner: Hostile tone and body language, not responding to 'Good Morning,' pushing past me in the hall."

Several days later, Blakey convened a business meeting in the lab to announce the grievance procedure Hill had invoked against him. He had already told the group that in the future, routine x-rays of the bones would be taken and analyzed by staffers at Howard's medical school who, Hill recorded him as saying, are "better trained" than she — a "direct slap in the face to me," and untrue as well.

By then, Blakey had obtained $20,000 from GSA to buy x-ray equipment for Cobb lab. It now might never be used.

Blakey told the osteological technicians and assistants (OTs and OTAs) to ignore Hill's requests for further cleaning of the remains. Blakey confirmed this in a taut memo to Hill on May 23. He said that henceforth, all her requests for help should go to lab director Mack, not directly to her OTs and OTAs. He also stipulated:

"No documentation of the African Burial Ground or other materials of the project are to be removed from the Cobb Laboratory without [his or Mack's] approval . . . . All data are under the custodianship of the Project, whose director [Blakey] determines appropriate analysis, curation, and dissemination of research materials and results."

This memo to Hill had one other fraught command. Speed up your work. Finish it by August 31, the last day of her contract, "after which I anticipate no further opportunity for you to make these assessments."

He planned to fire Hill.

Late in August, Hill wrote Bonner again, saying that Blakey was loading her with work — more than she could handle, scientifically or personally — while cutting her hours and assistance from others. "Blatant harassment," she called it. But these complaints were never adjudicated: On August 31, Hill's contract ran out.

Blakey told her to leave.

"Michael evicted me," she said.

SANKOFA?

She said later that she asked, but Blakey never explained, why he let her go. It was a painful blow for Hill. It soon would turn out to have been a fateful decision for Hill and the ABG project.

"I left as quickly as I could arrange for a moving van," Hill said later. "I drove the truck myself — back to Alabama!"

Since then, she has worked as a contract anthropologist-osteologist. "I was employed immediately for a major site by the University of Georgia." She also has led projects in Alabama, Arkansas and Tennessee, and Mississippi. At one site in Northern Alabama, in 2005, she was busy disinterring 200 remains of an unknown people from burials that appeared to be at least 2,000 years old and perhaps much older. The site was then covered by a road.

IN 2001, HILL EARNED HER doctorate (PhD) at the University of Massachusetts, Amherst. Her dissertation topic: porotic hyperostosis, the hallmark of anemia that she had found on many ABG skulls and skeletons.

Several matters with Blakey remained outstanding: One was a critical scientific issue: the continuity of the osteological research. Hill had written detailed notes on each of the 373 sets of remains that she processed. What she had *not* had time to do was enter these findings into the ABG database, an important step that is called *coding*.

Hill said that having one person analyze the bones and a second code them into the database is scientifically unacceptable. To prevent "inter-observer error," she added, the same scientist needs to analyze the bones *and* code the information. If she was correct, then the chance of ever obtaining solid data from the ABG remains had been severely compromised, all the more so since the bones have now been reburied and thus can't be re-examined to resolve disagreements.

She said that Blakey told her later by phone, "We've taken care of that," regarding the coding. Another scientist did it. According to Hill, this data-entry scientist was neither her student nor her OT. Neither was he an osteologist at the time, Hill said later. He was a graduate student.

Anthropologist Jerry Rose, PhD, in Arkansas, tended to support Hill's views on who should do the coding:

For ancient bones, splitting the two tasks — recording and coding — is "a concern.... It's hard as hell to enter someone else's data!" he explained, in a 2004 interview in Fayetteville. "A lot is lost in the coding.... [It] is very subjective." * The verbal description has a subjective component. That has to be translated into the codes, which are definite, not nuanced. The task *can* be divided, he said, providing that both the diagnostician and the coder are on the same research team.

"In a sense," Rose added, "Hill is right, given how much they [Blakey and his supporters] are making of [the bones'] conditions."

Asked if, in his view, Hill is a meticulous scientist, and a good one, Rose answered "yes" to both questions.

Hill doubted that Blakey understood the problem he was getting into by cutting her loose halfway. "Michael had no experience in how to do this kind of analysis," she said later. "Neither did Mark. [Neither] of them had ever been in charge of anything like this."

Asked later why she had hung on in the Cobb lab for so

---

* This problem was well understood by Blakey; he and several remaining co-workers note in their Final Report: " For consistency, the same Osteologist [i.e., Hill] carried out most of these assessments. Where staffing changes were made for pathology assessment and coding, care was taken ... to establish comparability among researchers." Hill later protested that her replacement "is not qualified to code my data. He worked in the lab when I was there ... but deliberately ignored my instructions.... [He] has [n]ever consulted with me about how I would code the data."

long, Hill replied:

"Was I waiting for a miracle to happen? I guess I was! .. .. Hope springs eternal!" She added: "I keep thinking this project can be brought to a successful conclusion — it's such a major data set!"

Looking back on the occasions when Blakey harangued her to tears, Hill said that during these episodes a curious smile would appear on his face. Then, he would start discussing lab business in a calm tone, without reference to the shouting and weeping that had just passed.

IN UPDATE (WINTER 1996), Blakey reported that 365 of the remains had been cleaned, reconstructed, and inventoried. All 427 may be done by March, he said. By summer, "the project will be fully engaged in statistical research and writing of our first major report."

He urged his readers not to be intimidated by the word or concept of *statistics*:

"Statistical and demographic methods provide means of demonstrating group trends in carefully qualified ways ... [we] will give some ... group trends in the ABG population that tell just how widespread ... were the problems enslaved Africans faced.... The project reaches a new plateau as we begin the quantitative analysis of the biological data."

Blakey sounded upbeat. "As I lecture at the leading universities in the U.S.," he said, "I find that the ABG project's significance has evolved. Two years ago, the negative impact of racism and [the] heroic struggle of the African/African-American descendants in New York took center stage. The opponents of Howard University's approach to publicly engaged, biocultural, and anti-racist science also insinuated their objections more often than not. Over the last few months there has been a change in perception as people learn of what we are actually accomplishing here, and why. *This project has now emerged as the leading model for the way such anthropological research should be conducted as we approach the 21st Century"* [emphasis added].

Update's Spring 1996 issue followed up, saying, "the focus has shifted to the scientific and historical data forthcoming from Howard.... Blakey... expects to be able to present an evaluation of diet, lifestyle, even the African origins of this burial population within a short time." In other words, *Update* was promising results, soon, on research that had barely begun (genetics) and that had not been authorized or funded by GSA (genetics, diet).

In an *Update* (Fall 1996), published after Hill had left, there is a photo of the Cobb lab staff — 18 people. In it, she stands directly in front of Blakey, who towers over her by a foot. Both look dour. Hill is one of the few, if not the only white person in the picture.

Several months later, in the Winter 1996-97 issue, an *Update* editor said to Blakey, in an interview:

"Your research methods seem to be creating new standards for the future."

Blakey replied:

"Just about everything that we do has that potential.... Thus far, we are succeeding in just about everything we've planned and have begun to do."

Update had little to report on scientific progress in its August/September 1997 issue. It did say, however, that Blakey had been invited to a conference on human rights in Geneva, Switzerland. He spoke there on August 19th, on the ABG project's significance in blacks' struggle. He said, "the scientific research now under way" constitutes a new dimension in the "long-standing human rights struggle among African Americans." Blakey was appointed a Permanent Representative to an educational unit of the Organization of African Unity.

SANKOFA?

At year's end, 1997, the December *Update* carried a report by a representative of the Admiral Family (see p. 65), the Inman Muhammad Hatim, on Blakey's speech in Switzerland. The audience "was held spell-bound" by Blakey's discussion of the ABG, the Inman reported. But, he added, the Admiral Family continued to worry, "What are we going to do about the remains of our ancestors?"

The bones now had been above ground for five years — and a reburial date, while long promised, had not been firmly set.

There was no new news from Cobb lab in the Winter 1997 *Update*. Nor was there any in the Spring 1998 issue. Neither was there in the summer one. But, in fact, much was happening.

Credit Where Due? After this feature appeared in *National Geographic* in October 1997, Cassandra Hill wrote to the magazine saying, "I was dismayed to see that you . . . erroneously appear to give full credit to Dr. Michael Blakey for the skeletal analysis. [He] is the project director, but th[is] . . . . is a multidisciplinary and multicultural project . . . . [When] I was the osteologist . . . I was solely responsible for all skeletal pathology analyses, including that of the child to which you refer . . . . It is a disservice to all those involved in the research to present such a singular credit . . . . "

Geographic did not publish Hill's letter. But the writer sent Hill a note saying there was "very little space . . . to detail the many contributors and their valuable efforts. The article simply names Dr. Blakey, as he is the project director . . . ."

DAVID ZIMMERMAN

16

DATA AND DISCORD

ON AUGUST 5, 1998, MICHAEL BLAKEY submitted a "first draft" of the Skeletal Biology Report to the General Services Administration (GSA) at its Northeast regional office in Manhattan. He also sent copies to the project's Office of Public Education and Interpretation (OPEI) at the World Trade Center in New York, so that the report would "be available to research colleagues — and others."

It was a massive presentation. The Report itself ran 190 double-spaced pages. With it were two thick, unpaginated volumes of Case Summaries. Also: Two thick volumes of Appendices (also unpaginated). The whole package included some 2,000 pages.

There was one thing Blakey did *not* do before submitting the draft Report: He did not submit it to Howard University for review, according to a close colleague, Howard University geneticist Rick Kittles, PhD (see Chapter 18). Like many academic and research centers, Howard maintained a committee to examine faculty manuscripts in order to forestall publications that might embarrass the institution.

The report's title page cites the ABG project at Howard's Cobb Laboratory as the submitter. A list of contributors, based on their titles, follows: Director Blakey is first. Mark Mack, listed as laboratory director and osteologist, is second. Next is the office manager and assistant to Blakey. She is followed by Cassandra Hill, osteologist. Blakey, Mack, and Hill are the *only* scientists listed. The names of 44 osteological technicians and assistants follow.

A list of "contributors to this volume" shows Blakey as sole or first author of 7 of the 11 chapters and secondary author of the other 4. Cassandra Hill is listed third among contributors to the chapter on skeletal pathology and epidemiology. A footnote says she "made large contributions to relevant data collection and/or analysis, but did not participate in the writing of chapters." The frontispiece is a photo of burial #315, an essentially complete skeleton, lying undisturbed in her grave after having been uncovered by archeologists working on site in New York City.

In an early chapter, the report says:

"Given [that] the sources of enslaved African captives were diverse, any attempt to treat the cemetery remains as a single group must be understood as potentially flawed. This population may collectively resemble groups to which they did not individually belong."

It is hard to know what Mack, who is the senior author of this chapter, meant here.

The report says:

"Given the high level of forced migration of enslaved

Africans into and out of the City, it is very difficult to tell from skeletal demography how may [sic] of these who died in New York would also have been born there, making estimates of fatality [sic] difficult to evaluate from the mortuary record alone."

This, too, is confusing.

Blakey and his co-authors revealed one important fact: The "forced migration" — transience — of black people into and out of New York means that the carefully determined *age at death* may provide an age profile of the cemetery's population, but not the community's. If, for example, older men moved or were sent elsewhere, where they eventually died, then a calculation that 50% of the burials were children could be meaningless in that a significant number of older people left before they died. If the oldsters had been counted, perhaps only 40%, or even 30%, of deaths in the community were children. This problem, the unfathomable transience of adults in the community, would haunt all the project's efforts to generalize population figures from the ABG remains to the colonial black community at large.

Straightforward information is equivocated in the report. For example:

"We *suspect, moreover*, that men are *more* often affected by arthritis in this population partly because *more* men *more* often live to older ages when arthritis occurs *more* prevalently [emphases added].

The report is difficult to summarize because, as Blakey allows, "no strong conclusions will be attempted prior to the availability" of data from the historical and archeological ABG teams at work elsewhere. This prompted one scientist who later reviewed the report for GSA to scrawl on the page: "How can you go onto other questions without bringing present data to its maximum [?]."

The 27 fungally infected sets of human remains are not included in the draft report. Thus, 400 sets of skeletal remains (427-27) are included. Of these, 373, or 93%, had been examined by Cassandra Hill.

While awaiting GSA's reaction to the draft report, Blakey had other urgent problems to attend to. He was late on his deadlines, according to which the final report should by now have been written and approved. And the remains should have been returned to New York and reburied. Blakey also was way over budget. But he was short of cash to carry on the work.

As of August 1997, GSA had authorized and appeared to have paid $5,162,000, about a third of which had gone to the *historical* and *archeological* arms of the ABG project. The archeologists were sorting and studying the *grave goods*, items buried with the bodies and other non-human artifacts, most of which were turning out to be shroud pins and coffin nails.

The cost thus had reached almost $13,000 per set of remains ($5,162,000 ÷ 400) — a prohibitive cost, Hill said.

Up to this point, the feds had been willingly writing the checks. But officials were growing more and more unhappy about the cost.

"They [Blakey and his associates] were continually asking for more money," consultant Jerry Rose recalled later. It was "discouraging that they were always asking for more money."

The expenditures, he and others pointed out, were — or should have been — capped at $4,800,000. This equals 2% of the construction cost of $240,000,000 for the office building — double the allowable amount under the National Historic Preservation Act to "mitigate" the Burial Grounds' destruction. The figure is supposed to include the design and construction of a monument — conceivably a building — to honor the ABG population and display the project's research findings. This work had barely begun.

GSA officials were worried. Even before Blakey sent them his first draft, they had met with the Howard researchers (May, 1998), to lower the boom: Blakey said they told him that the agency would continue to fund his work for the year. Then they would stop.*

GSA would not fund the follow-up genetic studies. Neither would it fund chemical testing of teeth and bones, which might provide insight into an individual's nutritional profile and, hence, perhaps also their natal origins. Moreover, GSA now planned to exert greater administrative control.

These changes were explained in an "Open Letter to ABG Friends" by a GSA administrator, Thurman M. Davis, Sr., which *Update* published on the front page of its Winter 1999 issue. Davis restated GSA's ongoing commitment to the project, then addressed the outstanding issues:

In *re* money: GSA's commitment in 1991 was for 1% of the cost of the 290 Broadway building — $480 million — to "implement data recovery and mitigation" due to the cemetery's destruction. GSA would "ensure that the research plan is implemented." In 1998 current and planned expenditures, including both research and "memorialization" of the site, were now pegged at $15 million — "nearly four times the 'not-to-exceed'

---

* "Your statement" that GSA will end its funding in April [1999] "is simply not accurate," the agency's area administrator, Robert W. Martin, angrily told Blakey. Martin said that Blakey lied when he claimed that GSA had failed to respond to his request for new lines of research. The problem, he said, was Blakey's:

"In light of GSA's consistent history of granting ... requests for extension, and your failure" for seven months "to provide ... further proposals or additional project[s] which we've indicated a willingness to fund, your statement about GSA's ending project funding in April 1999 is an especially egregious misrepresentation."

amount stipulated" in the Memorandum of Agreement that Blakey had signed. This is "an extraordinary level of federal support," Davis wrote.

In re research at Howard: The present $5.2 million contract would end on April 30, 1999. At Howard's request, GSA already had granted 31 months of extension on what originally was a 36-month contract. GSA could continue to provide extensions but would not provide more money until after reburial — now projected for 2002.

Blakey, responded in the next (Spring 1999) *Update*:

"The 1% stipulation was over-ridden by GSA's subsequent agreements with the community and other agencies. . . . because the site is more important than average," he claimed. He added: "GSA agreed to do far more [than 1%] . . . in order to correct the problems it created by desecrating and destroying our oldest African Burial Ground."

Blakey cites no specific agreement for this added expenditure. But elsewhere, he had said that it is contained in the final amendments to the Research Design, in 1993, which, he claimed, obligated GSA to fund all research described in that document. (See p. 104.)

Contrary to this, the proposals in the Research Design for the DNA and dental chemistry analyses are neither specific nor clearly stated. They are vague. Blakey had written, for example, "we are interested in testing genetic affinities to African regional populations and potential cultural groups . . . ." The data base "involves" the sequencing of relevant extended haplotypes [DNA profiles]. *Involves* does not signify intent. The only indication of Blakey's research intention is the indirect statement, "We will proceed by assessing the combination of traits in each individual. . . ."

This certainly is a fragile basis for his claim that GSA had committed the government to pay for genetic analysis of the remains, at an additional cost that was then estimated to be approaching $5 million. The full statement on DNA analysis is reproduced below, exactly as it appears in the April 22, 1993, draft of the Research Design.*

This is in sharp contrast to Blakey's rebuttal to Davis that "GSA has understood . . . from the very beginning . . ." that the scientific research — including genetic and chemical analyses — "would cost at least $10 million." He cites no source for this "understood" promise.

The enhanced oversight that GSA now sought was manifest in the creation of a new administrative position, *Executive*

---

* "[G]enetic data from the African Burial Ground population will provide information concerning biological continuity and change resulting from African forced migration to North America. We are interested in testing:
  • Genetic affinities to African regional populations and potential cultural groups.
  • Effects of selection for/against particular genes as a result of adversities of The Middle Passage.
  • Characterizing baseline genetics from which African Americans evolved.
  "The data base for all these analyses involves the sequencing and electrophoresis of relevant extended haplotypes, including Rh, Gm, HLA (B-42) and extended B-globin, L-globin, mtDNA and Y chromosome regions (Y-chromosome data will also augment sex determination). Comparative data on living African and European populations are currently available, and bone samples will be obtained from 18th-century archeological populations if possible. . . . We will proceed by assessing the combination of traits in each individual, and probabilities for affinity to particular parental populations will be based on cluster analysis of the individual's trait combinations. These individuals will then be assigned to population groups for analysis, or as pan-mixed populations as the data warrant. . . .
  "DNA data will be compared with morphometric analyses (measurements of the form of the skull and face) which, although less precise than DNA as to a population's genetics, will allow comparisons with a much broader data base of archeological populations than have traditionally been studied on the basis of morphological variation. The biostatistical methods to be used conceptually operationalize the tenets of the phenetic approach to taxonomy which assumes that population similarity in numerous characters is indicative of genetic relationship."

Director of the ABG. An official named Lisa Wager was appointed. The GSA's real purpose, Blakey complained, is "to oversee us."

Wager is "a Euro-American who lacks the necessary expertise and understanding of the African American community," he wrote in the Spring 1999 *Update*. He accused GSA of sabotage:

"What [it] has done is equivalent to making an African-American who is not knowledgeable about Jewish life and history the head of the Holocaust Museum, and [then] aiding that person's continuous interference with what the Jewish community wants the museum to do. Would that be allowed? I don't think so[!]."

GSA backed down. In the summer a black political consultant, Ronald Law, was appointed as GSA's project manager for the ABG, replacing Lisa Wager. Later, in 2007, Wager declined to comment on the episode.

An *Update* editor, Emilyn Brown, interviewed Law, for the Summer 1999 issue, and asked: How do you answer critics who say your appointment is based on race?

Law replied:

"My appointment is not based on race, and I will not engage in any discourse with anyone who suggests that [it] was . . . ."

Law said his goal was "*closure*" [his emphasis]. We need a memorial, he explained, and reinterment of the bones and an educational center so that America "can learn about Africans in the early part of American history."

What must have weighed heavily on Blakey's mind in the spring of 1999, besides the project's emptying pockets, was GSA's reaction to the first draft of the skeletal biology report. It now had been in the agency's hands for many months, and its assessment was hard to discern. In May, Blakey wrote Lisa Wager, who was still on the job in Washington, to find out. Blakey told her he wanted to see "all reports of GSA's consultants, including... Dr. Jerry Rose, who have been evaluating our progress to date. As you know," he added, "I have not received those reports, and although you have said they were highly favorable, I think that we need to see the actual reports."

Meanwhile, Cassandra Hill, back in Alabama, had not forgotten her important role in what she, like Blakey, deemed the ABG project to be: one of the century's most important American anthropological finds — and its investigation. She had three principal concerns: Could she publish her findings? Would she receive authorial credit for her work on project publications, and third, would she be able to read the reports in draft in order to correct any misinterpretations? This would ensure that her name would be attached only to accurate information. Her attorney, Russell L. Sandidge, wrote twice in 1998 to Howard University attorney John A. Jackson raising these (and other) issues. A whole year went by before Jackson responded:

Hill would be listed as "a contributor" but not necessarily as an author. "Authorship," he wrote, "is seen as a function of writing" the report, not of having done the research. Hill, obviously, had not helped write the draft report.

Blakey might, however, consider her for a "co-authorship" if she would write a research paper on skull fractures, which had been her particular interest, then write a second paper on porotic hyperostosis, her area of expertise, and then write a third chapter on "pathologies in general." She couldn't have access to the skeletons at Howard. But she might view photographs of them elsewhere during the summer of 1999 —

which, however, was all but over by the time the Howard lawyer's letter reached her in August.

Any and all of her inclusions and credits would be at Blakey's discretion, Jackson stipulated. If she didn't like what the report said, she could remove her name from it. She "cannot" be given space for a separate critique of the main report.

Feeling she was being led on, Hill declined this offer: None of the other researchers had been required to write papers in order to be "considered" for authorship. But she was not reconciled to being cut out of the major scientific credit for the work she had done.

Denying her credit for her work — or, at best, demanding that she beg for it — deviated sharply from the Research Design, which stipulated that "the specialists who have the greatest technical expertise and responsibility for [the chapters'] content" will write the chapters. "Researchers involved in the design, analysis, and writing of these chapters . . . will be attributed authorship in accordance with their level of effort."

17

REJECTION

On August 12, 1999, one full year after Michael Blakey submitted it, the General Services Administration (GSA) wrote to him about the first draft of the skeletal biology report. GSA rejected it.

The agency based its decision on recommendations from internal scientific reviewers and external reviewers, including Jerry Rose. GSA rejected the draft as inadequate. As incomplete. Erroneous. Rife with errors — including many misspelled words. The 2000 pages contained lots of data but little text and essentially no analysis. Much of what was written was meandering commentary, not scientific information.

"It wasn't an analytical report!" Rose said later. It had "no substance."

Asked why it had taken so long for GSA to reply, Rose said he wasn't sure. "I guess," he ventured, "they were so overwhelmed that the job was poorly done that they didn't know what to do!"

This was the work about which Blakey had recently bragged, "we are succeeding in just about everything . . . ."

"This draft has many strong and weak points," GSA said. But, "the overall evaluation is [it] is not complete. . . . This cannot be considered a draft that will, with minor modifications, become the report of work conducted to date."

It took Blakey four months, until December 21, 1999, to provide a "detailed response" to GSA's rejection letter. He sent it to administrator William Lawson at the agency's regional headquarters in New York.*

Blakey's response was deferential:

"The reviewers are absolutely correct that the draft is like an extensive progress report," he said, and a "final report of the project as a whole cannot be completed with minor modifications." Some "weaknesses," he adds, result from the fact that "the report is a first draft which solicits review and feedback prior to the writing of a completed draft."

Blakey then parries:

"How is it," he asks, "that the major general criticism of the reviewers concerns characteristics . . . that we indicated were intended, i.e., that the report represents one step . . . toward a much larger and more integrated report."

Given the "newness" of the "comprehensive approach"

---

* This exchange of letters appears to be the most revealing documentation of what went wrong with the ABG Project. It has not been previously exposed. The Author was fortunate to obtain a copy of Blakey's rebuttal, and we all are fortunate that he cites, between quote marks, Lawson's rejection letter, so both sides can be reported here. GSA, however, over several years, flatly refused to provide copies of this interchange under the Freedom of Information Act (FOIA). Indeed, GSA mostly refused to provide any documents or information pertaining to its deteriorating relationship with Blakey or the cover-ups that both he and GSA mounted to hide the truth.

proposed in the Research Design," Blakey advises Lawson, "reviewers might not intuit that the processes of interdisciplinary reporting is [sic] fundamentally different from the usual archeological site reports." He explained that all the reports by the researchers elsewhere, who are studying the [nonhuman] archeological remains, and the historical record will need to be in before the "integrated" report could be written.

Blakey here, and elsewhere in his 10-page single-spaced rebuttal, appeared to confound criticism of his group's failure to analyze the data from the bones, which should already have been done, with production of a final, multidisciplinary report that would be written in the future, after all three individual reports — anthropology, archeology and history — were completed.

Then, Blakey passed the buck back to GSA for the draft's shortcomings:

"[T]he reviewers might not have been aware that the GSA impediments and delays of funding have slowed or even completely stalled the completion and start-up work in many disciplines," he said, meaning specifically that the DNA analysis and chemical analysis of the bones had not been funded. He then added, ingratiatingly, that his response to "this most overarching of criticisms" — that the draft is incomplete — was that the reviewers' comments "are consistent with the project's expectations and the intended purpose" of the rejected draft.

Getting down to the nitty-gritty — the first draft's particular shortcomings — the reviewers, said Lawson, found fault with the "minor spelling and grammatical errors."

These mistakes would not have surprised Cassandra Hill. She said later that she was not sent — and did not read — the first draft. When she learned years later about its reception by GSA, she said that she had warned Blakey and Mack that inexcusable errors would be held against them.

"When the reviewers get that, and they see you can't spell the bones of the body, they won't get past that!" she said she had told them.

The GSA asked Blakey and his colleagues to "summarize the major findings in relation to the questions [he] proposed to answer in the original 1993 Research Design."

Blakey replied in his letter to Lawson:

"We cannot do [it]."

Reason: GSA had not funded the DNA and chemical research. "[W]e will not attempt to make bricks without straw," Blakey said. But, "we have continued to work with GSA . . . and responsible public leadership to seek accountability of the Government to providing [sic] the means which in 1993 were agreed as needed . . . to answer the GSA's questions . . . ."

In other words: GSA was responsible for Blakey's failures!

"Every problem he ever had was our fault," commented later GSA official Peter Sneed, who was continually in the fray with him at the time.

In his rebuttal letter, Blakey promised to revise and resubmit the first draft, following the reviewers' recommendations. He didn't.

ONE COULD ASK: WHY, after all the effort, cost, and time did Blakey deliver so poorly? Didn't he know the first draft was inadequate? If he didn't know, he certainly had a poor understanding of what is required for scientific reporting. Or, was he caught up in his delusional sense of his own role, as creator of a new form of ethnically specific, "multi-disciplinary" science?

He seems to have consciously believed that the scientific

community supported his methods; after all, the anthropology department at UMass Amherst had approved his doctoral thesis. And in fact the rambling, desultory text of the ABG first draft is quite similar to that of his thesis, written a decade earlier.

Another, compatible possibility is that Blakey and his Cobb lab associates couldn't write what was required: They didn't know how. Not only did they have trouble with the writing, they may not have been able — may not have known how — to perform the required scientific analysis of the data. Cassandra Hill said much later, after examining the draft, that she could have — and would have — written it. But she hadn't been asked.

She further claimed that for the report to be valid, she should have done it. Reason: The analysis would have had to be based on her detailed, meticulous, hand-written data. If anyone but the primary osteologist were to use them, even in conjunction with re-examination of the bones, this would — and must have — introduced "inter-observer bias."

Without Hill, Blakey could easily have hired a freelance science writer to present the findings appropriately. He didn't.

What also is true is that virtually nobody knew about it: GSA's rejection of the draft was not reported in *Update*. It was not told to the media either by Blakey and his associates or by GSA. Each side, of course, had its own reason to be embarrassed: Blakey for the rejection. GSA for letting it happen.

The copies of the draft that Blakey had placed in the Office of Public Education and Interpretation (OPEI) library disappeared, and OPEI chief Sherrill Wilson, refused to provide one. The wider world may have sensed that, in some way, the project was in trouble. But when Blakey did discuss his problems, his issue was GSA's betrayal in not funding the DNA and bone trace mineral studies — not his own failures.

The project was at its nadir. Blakey dated his letter to GSA's Ronald Law as December 31, 1999 — the last day of the millennium. This also was the final deadline for completion of the ABG project. The GSA inspector general ordered an audit to find out which of the deliverables had been handed in — and which hadn't.

GSA told Blakey to lock the cabinets where the remains were kept. In February, the locks were changed at Cobb Laboratory and at the project's archeology lab in the 6 World Trade Center building in New York. GSA said Howard University had changed the Cobb lab lock, Blakey reported in *Update* (Winter 2000). With grant money no longer available for salaries, Blakey said, he and others "have had to commit to other jobs that pay the rent."

Later, summarizing the situation from GSA's point of view, administrator Sneed said: "My position was that Blakey did a lot of manipulation of GSA, taking an adversarial position instead of working with others."

In the rancorous verbal exchanges that followed, one key fact remained unsaid: GSA's rejection of the first draft report. Blakey continued to blame GSA's racially motivated stinginess for the crisis. In fact, *Update* never told its readers about this failure. GSA hinted at, but did not explicitly reveal that the work failed to pass muster.

GSA administrator Lawson summed matters up this way:

"What was there to show for the $5 million that was spent? That was GSA's biggest problem. There wasn't a lot there!"

This figure, what is more, does not include expenditures in time, materials, and money by GSA. The agency was spending five dollars or so on its own work for the project for every dollar that went to the research.

How reasonable were the research expenditures, compared to similar scientific projects? When asked, Ted Rathbun, said by phone that he excavated 37 sets of remains from the slave cemetery he had studied in the early 1980s. He estimated the cost of excavation as $6,000 to $8,000. If the cost was, say, $7,000 for the 37 sets of remains, the average cost was $360 each.

At the Cedar Grove Cemetery in Arkansas, Jerry Rose excavated 98 burials, at an average cost of $2,027 each, in 1980 dollars, "from in-the-ground to publication" of his scientific report, *Gone To A Better Land*.

A far larger graveyard, Freedman's Cemetery in Dallas, Texas was excavated between 1991 and 1994. A total of 1,157 sets of remains was recovered, according to biologist Keith Condon, PhD, at the Indiana University School of Medicine in Indianapolis. Condon, the lead author of the technical report on the research, and his associates in the Texas Department of Transportation said they spent $1,410,000 on this work, or $1,305 per set of remains.

For the ABG project, skipping any post-2000 costs, Blakey exhausted his federal contract for $5,436,351 in processing roughly 400 sets of remains. That's $13,590 for each set. Adding the overhead would probably raise this by several thousand dollars per set of remains, since Howard University received an additional overhead payment of 45% on wages and salaries.

Translating these cost figures for the years they principally were expended to 2012 constant dollars, using the Consumer Price Index, gives:

- Rathbun's $360 in 1982 dollars = $855 per set of remains.
- Rose's $2,027 in 1980 dollars = $5,642 per set of remains in 2012 dollars.
- Freedman's Cemetery's $1,305 in 1992 dollars = $2,133 per

set of remains in 2012 dollars.
- Blakey's $13,590 in 1994 dollars = $21,035 per set of remains in 2012 dollars.

The ABG project's high research cost, which must have exceeded $25,000 per set of remains once the ancillary expenses were included, thus seems excessive compared to the others for essentially the same work. Rathbun said several years ago: "It seems like a spectacularly excessive expenditure for the scientific part of it!"

What is more, through 2008, GSA's contract with John Milner Associates had reached $7.5 million — and counting — according to GSA. Commented Rose: It was an enormous amount of money for the "piddling little results."

While suggestive, these cost comparisons must be taken with a grain of salt. Some, for example, include investigators' salaries. Others do not. Freedman's Cemetery expenses included coffins and preparation of the remains for reburial. The ABG's as yet did not. There are other discrepancies. Nevertheless, Michael Blakey and his associates spent more — much more — in constant dollars to study each disinterred set of remains than did their predecessors.

18

Genetic Forebears

Oadvertersity did not stop: the genetics research. The reason was that geneticist Rick Kittles, PhD, had his own laboratory there and his own money. He had come to Howard from the National Human Genome Research Institute (NHGRI), an arm of the National Institutes of Health (NIH), in nearby Bethesda, Maryland. He brought with him an unrestricted $50,000 research grant, part of which he invested in efforts to trace African Burial Ground (ABG) denizens' and other black Americans' lineages.

NHGRI was subsidizing research groups, called National Human Genome Centers, at sites around the country. Kittles and Howard's chief of molecular biology, Georgia G. Dunston, PhD, were co-directors of the one at Howard. By New Year's 2000, when Cobb lab was shut down, Kittles had performed DNA tests on 40 of the ABG sets of remains. Kittles also was doing something his Howard colleague Matthew George, had not done; he was tracing ABG males' *paternal* ancestry using the Y, or male, chromosome. A Y is passed exclusively through the male line — father to son — just as mitochondrial DNA (mtDNA) is passed exclusively through the female line, mother to daughter. By studying the Y chromosome DNA, Kittles could trace a man's lineage to his past and distant male ancestors.

In a scientific presentation the previous year (1999), Kittles had reported that "[R]ecently, we have successfully genotyped several ... loci" from many of the ABG bone samples. "These [genetic] markers," Kittles said, "will continue to be typed in order to evaluate potential genetic contributions to health and disease."

But, he added:

"To date, data on the biologic history of enslaved Africans is very limited."

His study, nevertheless, represents "a vital link" in connecting black people in the Americas to their African origins.

Several months later, a Washington TV reporter, Sam Ford, learned of Kittles' work and interviewed him. "I knew black people would be interested, so I suggested he do the test on me, and I would do a story on what he found," Ford said.

Based on his DNA, Kittles told Ford that his father's people had links to Nigeria; there also were genetic markers suggesting that his mother's people "came from what we now call Somalia, Ethiopia, Niger, and Guinea."

Ford's interview with Kittles lit up the switchboard at Howard with calls from black Americans eager for him to trace their roots to Africa. Many other reporters spoke to Kittles as well: When their stories appeared, the calls multiplied. Through Kittles' work, the ABG project had touched a raw nerve in many black people:

They knew little or nothing about their distant antecedents. They wanted the genetic information, even though Kittles and other geneticists warned that they had too few comparable African samples. What is more, while Kittles might find an affinity between a black person and his or her precursors in an area of Africa, there was no way — at least as yet — to trace one's lineage back to a specific African village or family — which is what Alex Haley had claimed, falsely, to have done in *Roots*.

"This wasn't clear-cut proof of anything," said TV newsman Ford. "But having covered trials where DNA is used in the courts, it was acceptable for me. It was the best evidence we['ve] got of where my family might have come from."

A civil engineer in Memphis, Tennessee, Melvin Collier, said that using Kittles' method to determine his ancestors' place of origin would sharpen his sense of belonging during a forthcoming trip to Africa.

In Bonita, California, neurologist William T. Chapman, MD, had a similar goal:

"I know that if I go, say, to Cameroon, I'm not going to have anything in common with those people," he said. "I don't think I'm going to have a spiritual event. I don't know. I'm just curious!"

Finding out, however; could be a downer. Kittles told a *Boston Globe* reporter about a man who showed up in his Howard office wearing colorful African clothing. He boasted of his proud heritage as the descendant of a Mandinkra warrior, like the ones he'd seen in TV or movie presentations of *Roots*.

"I did the test," Kittles said. "And I said, 'Man, your [paternal] Y chromosome goes back to Germany!'"

SANKOFA?

As, in fact, did Kittles' own Y chromosome, when he tested it! This was a noteworthy finding, given that his skin is quite dark. His mother's lineage goes back to the Yoruba people in Africa.

Kittles had come up with a major revelation:

It was possible to link skeletal remains in the ABG to regions and peoples in Africa. It also was possible — and much easier — to link *contemporary* black Americans to their African forebears in much the same way: compare mtDNA from a black woman in Florida or New York with contemporary African mtDNA profiles, and what have you done? You've told the contemporary American something about whence she came! Ditto for black men. In this way, you have partially bridged, as it were, a genealogy that was ruptured by the Middle Passage and slavery. The DNA research, Blakey agreed, could ultimately prove healing for Afro-Americans and "help restore the specifics of identity that were deliberately damaged by slaveholders in order to make enslaved Africans less human."*

"We have no connection to where we came from," added Bruce Jackson, PhD, a molecular biologist who operated a genetic search program similar to Kittles' at Massachusetts Bay Community College near Boston. "That is the greatest travesty... we have to face....

"It really has to end!"

Kittles' approach trumped Blakey's. Blakey aimed to restore black Americans' ties to their African forebears by link-

---

* Slavers stole their captives' identities even before they left Africa: Christian churches were built at the coastal fortresses in which the men and woman were held, awaiting a ship that would carry them west. In those churches, they were forcibly baptized with new, Christian names, albeit they later well might be forbidden to worship in white American churches.

ing the ABG individuals' population, genetically, to contemporary Africans. But — strangely — he proposed no direct link between these contemporary black Americans and the ABG population; this link is only symbolic. Why Blakey did not seek direct links between ABG individuals' DNA and Descendant Community members' DNA remains unclear — and seems to me to be a major oversight. Kittles, to the contrary, starts with living black peoples' DNA and attempts to link it directly to their African ancestors, with no detour through the New York ABG. This is a far more direct plan — and it raises the interesting possibility of identifying present-day New Yorkers whose direct ancestors were buried in the ABG.

ONE NEW PIECE OF information about the remains came to light early in 2000. A recount showed that the study included 408 sets of remains, not 427 or 390 or any of the other numbers that had been bruited about. The reason there were fewer than previously reported, Blakey said, was that some individuals who had been photographed in their graves had not been disinterred for study. Also: Some grave shafts were empty. Because of these different counts, this report has used the figure 400 throughout.

PERSONAL CONTACTS AMONG THE ABG stakeholders appear to have been brief and rancorous throughout 2000. Blakey told the General Service Administration (GSA), in writing, that the scientific work on the skeletal remains was substantially complete. He also indicated the dates when he planned to finish the work.

Like Blakey, GSA also was under the gun: The Descendant Community was agitating for reburial of the bones and the building of a suitable memorial for them. But these steps couldn't be taken if Blakey's research caused further delay. "Unfortunately," GSA official Ronald Law said, "any delay or lack of funding for the additional scientific research will have an adverse impact on [the] schedule."

For this reason, GSA accepted a $750,000 Howard University grant request — the start for a possible $6 million followup study — and asked the White House Office of Budget and Management to fund it. The request was denied. At year's end (2000), President Bill Clinton signed a law directing GSA to complete the work with federal building project funds and leftover money from 290 Broadway's construction. There would be no new allocations.

This raises the question: Who in the federal government was setting policy for GSA vis-à-vis the ABG project?

Rose said later that he thought these decisions were made at the highest level of GSA — or above it. This appears to be true, as shown by a new, high-level appointment to the project: She came from a far different social stratum than the GSA bureaucrats or the research scientists.

Her name is Cassandra Henderson. She is black. She was a tall, very attractive, and very sophisticated young woman. She also was a celebrity: an anchorwoman on CNN in Atlanta, Georgia, as well as an on-camera boxing commentator on ESPN2's Tuesday and "Friday Night Fights" telecasts.

Despite the prestige and perks of her work, however, Henderson said later she felt trapped in her TV jobs and sought change. She obtained a White House appointment as a public relations "specialist", a *consultant*, to GSA, tasked with bolstering the ABG's image and helping bring it to completion. Henderson moved to New York in September 2000. One of her first suggestions was to create an interactive ABG website. It soon was mounted.

Then, within months, President Clinton and the Democrats were voted out of national office. The Republicans who replaced them liked her work, she explained in interviews near her family home in Winterhaven, Florida, and they retained her during President George W. Bush's first term. Henderson applied herself to the larger tasks of planning ceremonies for the reburial of the remains and creating a suitable memorial for them on the original site — which looked barren and was surrounded by an industrial chain-link fence. For the interim, she arranged for flowers to be planted inside.

Geneticist Rick Kittles, meanwhile, was thriving.

The explosion of news accounts on his efforts to trace black Americans' lineages created national and international interest in his ABG work. Colleagues complained, however, that his findings were premature — based on small samples and still uncertain methods. The chief of research at the Armed Forces DNA identification laboratory, Thomas J. Parsons, PhD, told Carey Goldberg of the *New York Times* that a black American seeking to trace her mitochondrial DNA to the lineage of her great-great-great-great-great grandmother well might succeed. But in terms of the genes that determine her physical form and behavior, and perhaps even her mind — genes in the nucleus of each body cell — only 1 out of every 128 genes would have come from that ancestor. "[W]hat that means in terms of who you are: It's only a small portion of a potentially giant admixture," Parsons said.

Kittles replied, in an interview with the *Boston Globe*:

"To a lot of blacks, knowing a little bit of the story is important . . . . This will definitely contribute a lot to understanding the history of African Americans."

Acknowledging the criticism, he held his own:

"What I provide is lineages that are undiluted," he said,

in an interview posted on a black website. "I can trace back paternal lineages all the way back on the paternal side. I can also trace maternal lineages from mother to mother to mother before that."

Kittles was excited by the public response to this work. He wanted to press forward on identifying present-day Americans' lineages. The problem: money. The tests were costly.

"It is very expensive," he said. "This is not a hobby you can do with equipment in your garage."

The ABG project was no help. Blakey had no money and no mandate to study contemporary blacks' DNA. So Kittles let it be known that he was setting up a website for this purpose at Howard University and would charge about $300 for each set of tests.

Almost at once, complaints came from all over, including NHGRI at NIH. "We had some questions if it was a commercial venture," the institute's public relations woman, Cathy Yarbrough, told the *Los Angeles Times*. "We're not involved in that!"

Howard University hastened to assure NHGRI that none of its $50,000 grant to Kittles had been used for commercial purposes. Albeit, Kittle's commercial plan grew directly out of the grant work he'd done for the ABG effort.

The harshest critics, however, were Kittles' ABG colleagues, Michael Blakey, genetics researcher, and Fatimah Jackson, PhD.

"This controversy might give GSA fuel to interrupt our work," Blakey told the *Los Angeles Times*, apparently without revealing that the work had already been interrupted. Blakey accused Kittles of theft:

"Someone has taken someone else's half-baked cake and started selling slices to the public," he complained.

He added in the fall 2000 *Update* that "using DNA tests . . . to help living African-Americans find their . . . roots . . . is an ABG idea." But he cited no evidence that Kittles had stolen it from him.

Kittles retorted, pointing out that Blakey and Jackson were anthropologists, not geneticists. "There were no new ideas generated by ABG," he said. "One can read books and documents by [geneticists to find these ideas]. That is information that is available in the public domain."

Jackson objected, as quoted on the website salon.com, saying that charging blacks for DNA genealogical research was "like charging Holocaust victims to confirm their relatives were gassed." She declared: "There are certain things you don't charge people for. We're talking about American slavery, forced migration, prisoners of war. I don't think you ask the Descendant Community to pay for something that's their God-given right!"

Kittles' riposte: All services like his, of which there now were a growing number, charged for DNA testing.

The upshot was that Kittles left the ABG project. He said later in an interview that he "resigned"; his differences with Blakey were matters of "style". NHGRI public relations woman Yarbrough said later in a phone interview that Kittles "was fired."

The ABG's DNA data belonged to the project, so Kittles couldn't publish it — and it appears that it has never been published. Kittles took down his website. He stopped giving press interviews. Asked why, Howard University spokeswoman Donna Brock said: "His work is incomplete and has not been peer-reviewed."

In a report to *Update* (Fall 2000), Blakey declared that "no DNA testing service [is now] available" — albeit it well might be when the ABG project has finished its work. "It appears," he wrote, that Kittles' "premature service was quickly shut down in response to my inquiries with Howard University officials. [Some of them] say [it] was never formally started...."

Contradicting his assurances in the Research Design that mitochondrial DNA and Y-chromosome data "are currently available," Blakey now declared in *Update* that "the world's geneticists previously had not been much interested in the origins of the African diaspora," and so had collected very few data in places that sent people to the slave trade. Hence, the ABG project would have to go to Africa to do this — a major expense.

I FOUND NO EVIDENCE that either Blakey or Howard or GSA told reporters that the first draft of the skeletal biology report had been rejected. But advanced studies on genetics and bone chemistry probably couldn't go forward until the report was redone and accepted by GSA scientific reviewers. To say the least, all parties, by now, may have been highly embarrassed for the time and money wasted and for the unkept promises.

Blakey was worried that through Kittles and other ex-ABG project personnel, word of the research findings would reach colleagues and the public, outside of his control. In the mid-1990s, one of the early conservators on the ABG project, now long gone from it, had teamed up with a writer to produce a Young Adult book. The conservator, Gary S. McGowan, is white; the lead author, Joyce Hansen, is black. Their *Breaking Ground, Breaking Silence, the Story of New York's African Burial Ground* (New York: Henry Holt, 1998), is a clearly written and sympathetic account of blacks' lives and struggles and the effort to recapture these lives through the ABG research.

"The fully intact, well-preserved skeleton [Burial #6]

under the streets of Manhattan was one of the most important archeological discoveries of our lifetime," they say. Their book won a "Brotherhood Award" from Coretta Scott King, the late widow of Martin Luther King.

The book's writing is dispassionate but nevertheless quite moving; the facts tell the story. The illustrations are clear and carefully captioned. This book is one of a very few accounts published thus far in or as books that are free of rhetoric and special pleading. *Breaking Ground, Breaking Silence* carries the ABG story through about 1997. One other credible account stops in 1995. It is a chapter in *Unearthing Gotham, the Archaeology of New York City* (New Haven: Yale, 2001). The authors are archeologists Anne-Marie Cantwell, PhD, and Diane diZerega Wall, PhD.

Michael Blakey did not enjoy reading *Breaking Ground, Breaking Silence*. In *Update*, he calls it an "unfortunate book" and "an example of an unapproved and unauthorized publication which does not represent the work of the ABG project."

McGowan "did, however, use project resources for this unfortunate book when he had been employed to conserve artifacts from the cemetery."

McGowan, however, had pre-dated Blakey in working on the ABG remains. He said later, in a phone interview, that he'd never signed an exclusivity agreement with Blakey. "I think Blakey's problem with the book was that he felt nothing should be published that he wasn't involved with."

Meanwhile, the dust up between Blakey and Kittles had another unpleasant consequence for the project. Ever since the early disputes with the forensic anthropologists in New York City had ended in 1993, press coverage of ABG events had all been positive. Most of what was written or shown in the media was barely concealed public relations work, not independently gathered and weighed news. What mainstream reporter or media outlet would wish to look critically at black scientists searching for their roots!

The Kittles episode bared, for the first time, the intramural dissension among the researchers. The smell of blood piqued reporters' interest in the conflict and in the fact that the project was now long overdue. And the fact, too, that phone calls to Cobb Laboratory at Howard elicited this recorded message: "The research of the African Burial Ground Project has been interrupted for an indefinite period."

What had happened? A few reporters now wanted to know.

19

SECRETS

ONE OF THE FIRST thrusts came in the *New York Daily News* on February 5, 2001: Reporter Robert Ingrassia's story was headlined, "$21 Million Plan Mired in Woe — Researchers, Feds Wrangle over ABG."

Ingrassia wrote:

"Except for two wooden signs and a chain-link fence enclosing an easily overlooked plot of grass [next to the 290 Broadway federal office building], the government has almost nothing to show for its time and money."

Looking back, he said, "the project held great scientific potential and tremendous cultural value. Today it is mired in racial mistrust, funding feuds, and bickering . . . ."

The GSA, Ingrassia continued, is auditing Blakey's accounts to try and figure out how he spent more than $5 million. He quotes Blakey as blaming racism — the government's disrespect for black people — for cutting off the money; the feds are forestalling the DNA research that would help find the slaves' origins and permit living black Americans to trace their roots.

"What GSA has demonstrated," Blakey told the *News*, "is that it still wants black people to work for free!"

GSA sources, however, told the reporter that before cutting the money off, they had questioned why he was spending so much on administration and so little on actual research. They appear not to have been satisfied with his answer, according to GSA public relations specialist Cassandra Henderson, and so they ordered an audit.

Blakey wouldn't tell the *News* how much of the $5 million went into his pockets as salary or fees. He did say he never billed more than $50 an hour and billed for only about half of the hours he worked.

What neither GSA nor Blakey told the reporter directly was that the first draft of the skeletal report was unacceptable — and hadn't yet been brought up to par. Pressure, meanwhile, was mounting from the black community to rebury the remains. "We agree there should be research," Ingrassia quoted one of their leaders, Charles Bannon, as saying, "But come on, now, it's been nine years. . . .

"How would you feel if your parent died and some government [sic] took control over the remains!" Bannon said. "People are playing games with a deep spiritual ritual," he added. "This is the second killing of our ancestors!"

Henderson could not save herself from being caught up in these conflicts and the ill will they engendered. She became depressed, she and her husband said later. She found it hard to get up in the morning and go to work. She wanted to quit — but she didn't.

Henderson said she had little interchange with Blakey, but

what there was was unpleasant. One day, at a meeting, she recalled, she referred to him in conversation as "Michael". He told her that only his inner circle could refer to him familiarly in that way: For her, he was *Doctor* Blakey!

Henderson said later that she couldn't understand why he was so angry.

The GSA officials with whom she worked knew that the agency was being ripped off, Henderson said. But they were afraid to push back publicly, for fear their long-time complicity would lead their own heads to roll if, for instance, Congress were to hold an oversight hearing.

A second, related audit, finished later in 2000, assessed Howard University's handling of the costs. This report card was better: GSA had agreed to pay $5,436,351 for the work. But the University submitted only $4,575,346 in vouchers, leaving $860,987 unrequested and unexpended. The audit added that Howard "applied the proper labor and indirect expense billing rates in their voucher submissions to GSA."

This audit lent credence to Blakey's complaint that GSA was stiffing the researchers for money. Exactly why and how GSA could tell him, as it did, that the money had run out, when there seemed to remain a credit for the work of almost $1 million, is unclear. But it supports his belief that GSA, and perhaps also Howard University, were trying to starve him out of the project. No final financial audit was ever done. "It would take an entire shipload of auditors to figure it out," GSA's Peter Sneed said later.

When the GSA's second audit of the ABG project was completed and sent to administrator Ronald Law in May, it confirmed the bad news: Of the 13 documents that had been due the previous December 31, seven of them, including, of course, the Anthropology Final Report, were delinquent. The

Archaeology Final Report and the final historical report were also outstanding. GSA might have started administrative proceedings against Howard. But apparently a political decision was made not to do so.

The conflict between Blakey and GSA inflamed the Descendant Community. One New York City activist, in a letter to *Update* (Winter 2001), called it a "horse manure" game "to once again confuse African people." The purpose: to get Africans to distrust their hand-picked leaders, Adunni Oshupa Tabasi declared. "The U.S. doesn't want this history to get out!" Tabasi screamed at GSA administrator Law during a public meeting early in 2001. She told the black GSA official: "You're just a messenger boy carrying out your master's orders!"

Law was caught in the middle between the Descendant Community, which saw him as an Uncle Tom — a sellout to whites — and his GSA colleagues, who feared that he might be soft on the descendants. His aim, he insisted, was to remain objective, get the work done, and rebury the remains.

The Republican resurgence in Congress in 2000 created political pressure, and 9/11 increased the urgency to get it done, Law said in a 2012 interview in Lower Manhattan. "The Republicans said, 'We're going to shut this down!'"

Blakey would have none of this. "Michael wanted to expand the scope of the work," Law said. He "kept wanting to do more research!"

Blakey rationalized the criticism of his rejected draft report by insisting that the additional studies he proposed would make the work whole and allow its completion. Administrator Law, while opposed, seems to have accepted Blakey's view of their impasse.

On March 10, 2001, GSA held a public meeting to clarify

the project's status. The next day, Blakey was quoted in *Newsday*, a local daily paper, as saying:

"As far as I am concerned, there is no commitment or encouragement from the GSA to fund us."

A reinterment date of August 17, five months later, was announced. It was soon deferred.

Then came 9/11. The ABG's Office of Public Education and Interpretation (OPEI), on the second floor of the 6 World Trade Center — one of the Twin Towers — was totally destroyed. No ABG personnel were killed or injured. But the building's collapse also buried a significant number of the ABG archeological artifacts, which had been stored in the basement. At first, it was feared that all had been lost. But much of the collection later was rediscovered, in the basement, essentially intact, and was removed from the wreckage of the tower.

Through all the problems and wrangling, Blakey and GSA seemed to have agreed implicitly on one thing: As far as informing the public, silence was golden. Silence protected Blakey from the revelation that his scientific work was faulty. It protected GSA from the revelation that it had paid — more than generously — for contractual work that had not been delivered. Reporters' questions were shunned. GSA administrator Ronald Law, did grant this writer an interview early in the spring of 2002. He allowed that the peer reviewers for Blakey's first draft had identified things that still needed to be done.

"Howard is the lead university for completing the scientific research that is ongoing!" he said. Law explained that Howard had originally said it would need $5 million. "Now, they are saying they need more."

Howard needed to justify this added request, he said. And

the question remained: Where are the results of what they've done so far?

Administrator Law was amenable to Blakey's insistence that some additional research was "reasonable". But, "Is that the role of GSA? . . . I said GSA was not a museum!" Law added in retrospect, "He was pushing too hard. The information he had should have been sufficient — instead of going from A to Z!"

Three years had passed since the first draft submission. At a meeting the previous week, Law noted, Blakey had agreed to provide the agency with an estimated time for delivering successive drafts. Would Law provide a copy of the first draft?

No, he told me. It was a progress report and not releasable — even though a copy had been deposited in the OPEI reading room in 1998, and Blakey had invited the public to read it. (According to one Freedom of Information Act expert, once a document is released to the public, it ordinarily cannot be withdrawn. But GSA was not willing to release this one.)

GSA would provide nothing, Law said, until they had an approved final report in hand. He deflected all questions about what had gone wrong and why the target dates for finishing the project had not been met.

"We're making every effort to see that it goes forward," he said.

The ABG's director of OPEI, Sherrill Wilson, PhD, was similarly reticent. When faxed a set of questions intended to elicit the project's current status, she replied by phone:

"There is not one of those questions that is in my purview in public information."

Wilson explained that she knew only what she was told; her job was to provide information to interested teachers and school children. GSA had given OPEI some $5 million, through the contract archaeology firm John Milner Associates, to provide these answers — and nonanswers. Wilson suggested contacting GSA and Howard regarding these questions.

Blakey did not return several phone calls.

Six months later, in November 2001, GSA public relations woman Henderson said much the same thing as Law and Wilson:

"We're focusing on the future!"

GSA wouldn't say anything negative about any of the project participants, she said. She did say that the reburials could not take place until after the contracted reports from Blakey and others had been accepted. This is not GSA's decision, she explained. It's the decision of the Advisory Council on Historic Preservation. Henderson also said that GSA was still reviewing new contracts to complete the work. Call her back early in December, she said.

"We're not looking back," Henderson added. "We're looking forward to the reinterment!"

Nevertheless, the year 2001 had seen little, if any, progress toward that long-sought "closure".

SANKOFA?

20

GSA TAKES OVER

BY EARLY 2002, THE General Services Administration's (GSA's) aim had changed. Now it wanted to seize control of the project and complete it expeditiously. But help was needed. Since GSA still lacked qualified science administrators, the search for help focused on other government agencies that had the needed scientific expertise.

GSA looked, too, for a new technical advisor to replace Jerry Rose, who by now was dismayed by the conflicts and delays. He had served on a come-when-called basis. Now, he had become remote from the project. He was not prepared nor empowered to take a more direct role. Moreover, Rose's own research focus had shifted: He was spending long summer vacations away from his university, in deserts in the Middle East, in the Kingdom of Jordan and neighboring countries. He was exhuming Christian crusader burials, which, he remarked, the local Arab authorities — Moslems — were glad to see go!

So, Rose resigned his African Burial Ground (ABG) consultantship.

Later, looking back, Rose said Blakey "never asked for advice. [He] never thought he was in trouble." Asked if Blakey requested help on the preliminary draft, Rose added: "Nope! I never talked to him about [it]."

To replace Rose, GSA engaged a prominent forensic anthropologist, a government official: His name is Michael K. (Sonny) Trimble, PhD, a civilian employee of the US Army. Trimble, who is white, commanded the Army Corps of Engineers (ACE) center for the curation and management of archeological collections, in St. Louis, Missouri. This center's bland name belies its difficult and sensitive work. One of its major tasks is to search for and recover the remains of American servicemen and women lost in combat overseas.

(Two years later, in 2004, Trimble was widely shown on US television. He was in Iraq, where he directed the excavation of mass graves of people executed by the regime of Iraqi dictator Saddam Hussein. These macabre findings — thousands upon thousands of violated bodies — were sought as evidence for Hussein's trial by the new Iraqi government.)

The shift from Rose to Trimble — from *on-call* to *in-command* — came at an extraordinary meeting at Howard University on March 12, 2002. The meeting included all the governmental and institutional stakeholders, together with Michael Blakey and Mark Mack. The Descendant Community was not represented.

Besides GSA, there were representatives from the Advisory Council on Historic Preservation (ACHP) and the New York City Landmarks Preservation Commission (LPC); the National Park Service (NPS), which might manage the ABG

SANKOFA?

memorial once it was built, also joined the mix. But Trimble and his ACE team were clearly taking charge.

They set as their first task the convening of meetings with all key participants in the ABG work, followed by an evaluation of what had been done thus far and what was left to do. There also was a lot of fence-mending to do, since GSA had alienated officials from ACHP and LPC, whose advice it had shunned in selecting Blakey to head the project. The ACHP, for example, was willing to support the proposal to move forward, "provided" ACE would assume responsibility for "preparation of the [scientific] reports to ensure that [they] are finally completed."

In the same vein, the LPC wanted to be "assured" that data from the human remains and artifacts would be "extensive enough to prepare a quality report.

"In addition, LPC is concerned that the data be maintained in a format that will permit researchers to use ... [it] in the future, since further hands-on of the artifacts and all of the human remains, which will be reinterred, will not be possible."

Reburial before completion of the reports, and without the opportunity for other researchers to see and study the bones, would, of course, violate Blakey and Howard's fervent promise, at the outset, that other scientists would have access to these materials.

Trimble introduced two features that had been sorely lacking: *liaison*, which his team provided through face-to-face and telephone conferencing, and *communications*, which kept all parties informed and in forward motion. The communications came in an on line newsletter, called, simply, *Activity Report*, or, alternatively, the *Quarterly Report*, which was produced by ACHP every 90 days; the first issue, covering the

March meeting, is dated July 31, 2002. Besides its value for the participating agencies and scientists, *Activity Report* was to prove invaluable in my effort here to retell the ABG project's closing phases.

Before the March 12th meeting, GSA had told Trimble's group that Cassandra Hill's work was unacceptable. Blakey reiterated this view, the first *Activity Report* said:

"The Corps team began the project with an understanding that GSA felt the measurements of the skeletal remains were not sound, based on Dr. Blakey's statements during the ... meeting at Howard University."

But Blakey and GSA may have been trying to pass the buck for their own failures, for the *Activity Report* continued:

"However, after reading the peer review comments on the [first] draft Anthropology report, [and] reviewing selected sections of the data, and discussing the project with Dr. Rose, Mr. Mack, and Dr. Blakey, the Corps team came to the conclusion that the measurements are acceptable and do not require a complete set of new measurements.

"It is the belief of the Corps team that the peer reviewers ... feel the measurements [by Hill] are sound, but think the conclusions drawn from those measurements [by Blakey] need to be reevaluated and amplified upon." The ACE team went on to suggest that Blakey hire a professional editor.

After a follow-up meeting with Mack and Blakey, the *Activity Report* continued, "it was the opinion of the Corps team that they have gathered sufficient baseline data for the Anthropology Final Report." Cassandra Hill had gathered the lion's share of these data. Several weeks later, members of the ACE team revisited Howard and reviewed Hill's work. "The forms in each burial folder appear to be extremely thorough," they reported. So the ACE assessment here trumped Blakey's complaints about her work in the previous three years.

Mack's dental analysis remained undone, the ACE team said. They added that a "thorough report is absolutely essential." A deadline was set for the following December, but the team conceded that the work might come in a bit late.

Blakey told the ACE team that he had "a number of concerns about the peer review of the [still unwritten] draft final report.... He [Blakey] would like to suggest people to be part of the peer review of the final draft report." In other words, this time around, Blakey wanted to help pick the judges — the scientific peers — who would evaluate his work. A few months later, a high official in the NPS, archeologist Frank McManamon, PhD, volunteered to be a peer reviewer for the Final Report.

The ACE team that initially talked to Blakey and Mack concluded that their meetings were "successful," adding that "the Corps team believes that positive relationships are being renewed."

It was an optimistic view.

21

Betrayal

A DRAMATIC CHANGE OF direction was provoked in the summer of 2002 by a Page One story in the *Washington Post* (August 27). It was written by reporter Marcia Slocum Greene, and the key news, deep within it, was her apparently eyewitness account of an extraordinary meeting several months earlier:

"In May this year," she wrote, setting the tone for what was to follow, "the burial ground researchers retreated to Virginia's historic Moton Conference Center on the banks of the York River to assess the damage to their mission, and to salvage what they could . . . . [T]hey used the tranquil spot to confront what they believed had been the fundamental issue: race."

For years, Greene wrote, General Services Administration (GSA) officials had been unable to quell the researchers' "accusations" that their disagreements "were rooted in a racist

SANKOFA?

contempt for the advocates and the project."

She continued:

"Seated in large white wooden rocking chairs on the center's porch, [Michael] Blakey and archeologists Warren Perry, PhD, and Jean Howson, PhD, [who had been working on the cemetery's nonhuman remains] argued that the government's dogged insistence on a December deadline for the research left them without adequate time and money to complete the critical analyses. Research gaps in the final reports, the scientists feared, would make the team appear incompetent.

"'There is a gun to our heads,' said Blakey .... 'What GSA is insisting on is for us to fail .... I think racism plays a special part, as well as arrogance. GSA has demonstrated from the beginning a pattern of disrespect and disregard for the expertise of black people.'

"'I absolutely agree,' Perry said. 'These are our ancestors, our people ....' He stopped speaking, and tears filled his eyes.

"'They just interacted with us as though we were idiots and don't know what we are doing,' said Howson, who is white. Then she, too, began to wipe away tears.

"'We are all PhDs with years and decades of experience in this work, and their way of dealing with us on a day-to-day basis is that we're uppity and we're in their way and we're making their life miserable,' she said."

The story also contains a quote from Blakey referring to the $5 million thus far spent or committed for his report and the roughly $21 million total for the entire operation, including $56,700 for 420 coffins hand-carved in Ghana and $65,000 for seven mahogany burial crypts made in New York. Blakey said:

"We did not need what is often called, in black college circles, 'a colored grant.'"

Reporter Greene then quoted a former high GSA official, Brian A. Jackson, a black man, who labeled the racial accusations "hogwash". He added that GSA remains committed:

"This project is of tremendous scientific, historic, cultural, and community significance." For GSA "[t]o try to pretend that it is not, or to try not to address it would be crazy."*

Howard University officials may not have known the story was coming; there is no indication in it that reporter Greene had phoned to ask for their comment as the project's institutional sponsor. But when these officials read her story in their morning newspapers, they may have exploded in wrath, for Howard is, after all, one of those "black colleges" that Blakey speaks of in a belittling way, and like its sisterly educational institutions, it relies on so-called "colored grants" and other federal funds to sustain its annual budget.

"Howard was embarrassed," one knowledgeable GSA source says. "They didn't want anything to do with him!"

Blakey left Howard. Even years later, the school refused to say whether he had quit or been fired. He moved first to Brown University, where he stayed for the year 2000 as a visiting professor of anthropology. He later moved to the College of William & Mary, in Charlottesville, Virginia, where he became a professor of anthropology and the head of the school's new Institute for Historical Biology. He remains there today.

According to *Activity Report*, Howard wrote GSA "stating their unwillingness to complete the entire research project. [It] sought to withdraw from the agreement to complete the bioskeletal [and] archaeology report." University officials re-

---

* Ironically, while holding other writers, including this one, at bay, GSA officials in Washington and New York gave reporter Greene exclusive, wide-open access to the project, GSA public relations spokeswoman Cassandra Henderson said later. Why is unclear. Henderson said that she protested, but to no avail. So the agency was blind-sided by Greene's story.

iterated their wish to withdraw again, in a September 17 letter — all the while ducking GSA's many efforts to reach them.

GSA reacted in two ways: It opened discussions with the University of Maryland, in College Park, to see whether that university would complete the project if Howard failed to. And GSA wrote Howard a cure letter saying that its request to withdraw was "unacceptable". The very tough letter went on to state that "the University was endangering the performance of the contract, and that unless this condition was cured within 10 business days, the GSA might terminate the entire contract for default."

If Howard were forced to return the millions already received, or if it also were penalized further, the school's finances might severely damaged. These negotiations were conducted at "the very highest levels" of the two organizations, according to GSA's Peter Sneed.

Howard bit the bullet. A month later, the GSA Administrator Stephen Perry and a retinue of his aids met Howard president H. Patrick Swygert and his associates in the president's office. "During the course of the meeting," the Advisory Council on Historic Preservation reported, "Dr. Swygert announced that Howard was now fully committed" to finishing the work.

In going forward again, Swygert did what Howard had failed to do successfully the first time around: He set up a chain of command with two high university administrators to oversee the scientists' efforts. Blakey, Perry and a third researcher were to complete the bioanthropology, archaeology, and history reports, respectively. But Swygert's chief of staff, psychologist O. Jackson Cole, Jr., PhD, would be Howard's executive in charge of the African Burial Ground (ABG) project. And mathematician James A. Donaldson, PhD, would serve as project manager.

Blakey would remain responsible for the bioanthropology study — but now would need to report upward through, and satisfy, Howard's officials. In New York, meanwhile, Ronald Law left GSA.

Summing up the new directions, the July 11, 2002 *Activity Report* says:

"Dr. Trimble has met with representatives of Howard [and the stakeholder agencies] to bring the project back on track, to determine a project timeline, to hold stakeholders accountable, to gather information on where each party stands with relation to the ABG project, and to determine and solve budgetary, procedural, and public interest issues for GSA."

THERE WAS, AGAIN, ANOTHER pressure — and it was increasing. It came from the Descendant Community. Having been convinced, largely by Blakey, that GSA's endeavor was racist and hostile, the community now angrily blamed GSA for the delay, unaware that a major cause was Blakey's failure to complete the research as promised.

"The issue of community displeasure," Schomburg Center research head Howard Dodson explained to an Associated Press writer in October, "has been with the project since day one." He added: "Those issues remain a part of people's consciousness and concern."

Trimble and his Army Corps of Engineers deputies now were discovering what GSA had long ago learned the hard way: Once the ABG project ran off its track, it took much patience and time to put it back on. The task immediately at hand was to sign new contracts locking all stakeholders into the work needed to arrive at reburial and completion of the project. GSA public relations consultant Cassandra Henderson had told me in November to phone after Thanksgiving for confirmation that the papers had been signed. I phoned. They hadn't been.

Only on January 10, 2003, were the contracts finally approved. The new work under these contracts started on January 13, when the Cobb Laboratory at Howard University was officially reopened. To facilitate the progress, GSA provided each of the three teams with additional computer equipment, including monitors, scanners, and software.

For GSA public relations specialist Henderson, caught up in the newly invigorated ABG process, the motivation driving the participants had become clear. As she explained later, in their eyes, this grand project was not mostly about studying and reburying the bones, or honoring these ancestors. It was — *mostly* — about *money*: For GSA officials, salaries and bonuses. For scientists, wages and expenses that easily could top half a million dollars in just a few years. For members of the Descendant Community and their supporters, ABG paychecks, cash handouts, and rewards for service.

"The public process," Henderson said later, "is just like a business!"

22

'COMMUNITY' VIEWS

> *The popular embrace of DNA genealogy speaks to the rising power of genetics to shape our sense of self. By conjuring a biologically based history, the tests forge a visceral connection to our ancestors that seems to allow us to transcend our own lives. But will genetic identity undermine our cultural identity? The tests can add depth to what we have long believed, but they can also challenge our conception of who we are. The trauma some experience when their tests conflict with what they have always believed to be true has prompted some researchers to call for counseling to accompany the results.*
>
> — Amy Harmon, New York Times, January 22, 2006

RICK KITTLES, PHD, NO LONGER worked with Michael Blakey and the African Burial Ground (ABG) project. But the publicity his work had engendered, particularly in the Black Community, quickly turned the spotlight from Blakey's sequestered bones to Kittles' genetic studies.

He offered, perhaps, something different from the ABG

SANKOFA?

project: He offered black people news they could use in tracing their own personal ancestry back to Africa. This was — and is — exciting. It speaks to the deep sense of loss — of pain — for people whose forebears had been torn from their roots, leaving them stranded, unsure of themselves, in the hostile and hateful New World.

What could — or should — black people make of the new information that was or soon might be available to them, probably for a fee? Would it be helpful? Hurtful? As people once would have said: "Is it good for the 'race'?"

These questions were intensively explored at an unusual conference that was convened far from New York and Washington, at the University of Minnesota in Minneapolis, Minnesota, early in the summer of 2002. It brought black researchers together with community leaders to weigh — for better or for worse — the new methods that were beginning to shed scientific light on the black experience in America. The conference moderator was an attorney, john a. powell, JD (as he preferred to write his name), chief of the University's Institute of Race and Poverty. One of the sponsors: the school's Center for Bioethics, run by a physician, Steven Miles, MD. The gathering, he explained, is about this "disruption" in black Americans' lives and about "using the tools of history, genetics, and genealogy to examine" it.

Surprisingly, Blakey did not attend, albeit Miles said later that he had been invited. Neither, apparently, did any of Blakey's close associates, for none can be found on the participant lists. The ABG project was, however, present in spirit: Key black participants went on record to firmly oppose Blakey's mantra that the Descendant Community was *privileged*, over the ABG scientists themselves, in deciding what research was to be done, and how, and in interpreting the findings.

"There is a confusion about what science is supposed to be privileged for," explained evolutionary biologist Joseph A. Graves, Jr., PhD, of Arizona State University West, in Phoenix. "[S]cience deserves the highest privilege for ... explaining the way the natural world works. That is what it is designed to do, and does so well.... In other words, science will tell you how objects fall off the table and hit the earth."

However, Graves added, "There are other realms in which science should not be privileged. These include questions of ethics and spirituality. Science is not designed to be there. People confuse what science does do well ... and the other issues for which science has no role."

Blakey's perspective that the "community" shapes science thus was rebuked by attorney powell. He declared:

"Science is a discipline, and the Community does not participate in developing the protocols" — in this sense, the ground rules — "for science. This is not true just for science; it is true of every major [professional] discipline. They create their own language, their own rules ... and sometimes they do that without a great deal of input from the community. Even when the community has input ... it is as individuals rather than at the community level."

The conference's co-sponsor was a Black Community organization, the Powderhorn/Phillips Cultural Wellness Center, led by an African American, Atum Azzahir. Her group represents ordinary black men and women.

The conference, she said, is not about intellectual or academic matters. Rather, "it is about the community of African [American] people here in Minnesota who are attempting to bring together scholarship and living."

Key voices at the conference were black scientists — particularly Rick Kittles — as well as black intellectuals, scholars, and critics of biological and social facets of "race".

Slavery, attorney powell said, has been called "a kind of social death." It plucks people from their culture, their history, their language, and their identity. "If we try to reconstruct an identity — because we all need [one] — often times we are left with the tools, the symbols, the language, the religion, the customs, and the culture" of the slave-owning society and people who have caused that social death in the first place. "So it creates an incredible tension!

"[W]e are constantly co-creating . . . our identity.

"We are at a precipice," powell said — thanks to Rick Kittles. "[H]e brought us to this . . . place of having this dangerous opportunity to trace our genetic roots back to Africa, and back to particular tribes and particular populations."

Some of these risks then were elucidated by sociologist Troy Duster, PhD, of New York University. "I am ambivalent about using certain scientific procedures to construct genealogies," he declared. For one reason, these studies confirm what he and many other blacks already knew: that they are descended from slave *owners*, as well as from slaves. As Kittles had found, some 30 percent of black American men can be shown, through their DNA assays, to have had a white — and almost always male — antecedent. (In this way, Muhammed Ali, the black former heavy-weight boxing champ, has been hailed as an Irish descendant, through a great-grandparent on his father's side).

DNA evidence is now accepted as biologically "definitive," Duster went on to say. Thus, Thomas Jefferson has now been definitively shown to have fathered at least one of his slave — and companion — Sally Hemings's sons. Biologically definitive. But *not* socially definitive, Duster declared: Jefferson's white descendants refused at first to accept Hemings's black descendants into their kinship.

"No matter what the DNA says, Hemings' descendants

are not in the club!" But Duster had another, more serious fear about searching one's DNA for genealogic information — one that he was sure would be shared by the lower and middle-class black people in his audience: "the danger of embedding ... a racist formulation ... in one's analysis." In other words, DNA science may soon lead to the place where one can infer a person's ethnic origin — or what is called "race" — from the data.* This, then, is a double-edged sword.

The DNA technologies that are claimed to *transcend* race, in fact, on closer inspection, are "being deployed to recreate the very racial categories that we have said no longer exist!" Or, as Michael Blakey would have said, had he been present, Kittles' genetic assays could lead quickly to "racing". So, "[B]e alert!", Duster warned, because "inferring the ethnic origin of a crime stain," such as semen from a rape victim, can be dangerous, "because behind that is two hundred years of oppression."

"Here is our task," Duster counseled: Figure out "how to tease apart, analytically, politically, socially, these different uses. It would be dumb simply to say, let's get rid of the [DNA] technology. It would be equally dumb to say, Let's march forward! I'm suggesting that our task should be to try to dissect the different uses of the technologies, to applaud those in which we get people off of death row, and to be quite alert to and mindful of, those [uses] that reinforce old imageries about the nature and character of our 'ethnic origins.'"

Through such dangerous uses, what is more, not only might individuals or groups of people be harmed. But reification of the DNA findings could well re-create — definitively — the old and

---

*This situation appears to have changed late in 2005. On December 16, *Science* magazine published a ground-breaking report from an international team of researchers who claimed to have discovered a new gene — called SLC24A5 — that accounts for much, but not all, of the skin-color difference between black Africans and white Europeans.

SANKOFA?

now describedd "racial categories" under a new name: *ethnic origins*. African ethnic origins = black. European ethnic origins = white.

Rick Kittles from Howard University spoke next. Echoing Troy Duster, he agreed that "the challenge is to articulate the promise and the limitations of this exciting, complex, sensitive, and controversial" work — which is of particular interest to black people. In one web-based poll, he noted, 80 percent of black Americans "thought it would be important to use DNA testing to determine their ancestry."

Much shame is involved, too, Kittles said.

"I asked some . . . school kids about their African-American ancestry. Many . . . put their heads down; they are ashamed because they think about slavery . . . . It's embedded in our consciousness that our people started as slaves. . . . I work . . . to provide some information that would be useful for going beyond that wall."

Much of one's identity is based on one's parental name, Kittles added. "Genetics is not going to determine your identity," he said, "but it is going to help you shape your own identity."

How does "race" fit into this effort? Biologist Graves spent an hour explaining what "race" is *not*: It is not a biological concept that can be applied to the human race.

In scientific terms, *race* is biologically meaningful as a way of understanding closely related, yet still somewhat different members of a particular species — song sparrows, or red foxes, or honey bees, for example. But there is no racial division of humanity that makes either scientific or common sense.

In the 18th century, the time in which most ABG people lived and died, scientists disagreed on the specific traits that might define human subspecies, or races. Some believed that the differences that were found, including skin color, were due to environmental factors.

The concept of biologic human races arose in the 19th-century, Graves said. Not because of any change in what scientists understood about nature or human differences. "What really changed," he explained, "was the social importance of slavery and the social importance of declaring blacks, or the Negro, as a separate and inferior being."

Much has been written about the rank and rampant racism of some 19th-century scientists. But, Graves pointed out, Charles Darwin, in his *The Descent of Man*, doubted that races were real. Darwin noted that his racist colleagues could not agree on which human traits defined the different races. What is more, they were not looking at true geographical differences, which would have to have been a marker for "race", since the visible differences between people such as skin and hair color, in fact, do vary on a geographical basis. Blondes in Scandinavia, blacks in Africa, etc. But never exclusively. "That's why Darwin concluded that there was not consistent means to determine who belonged in what group — a conclusion that has not been changed by the huge volume of genetic data that have since been published.

"All the scientists [among the United Nations Educational, Scientific and Cultural Organization's authorities] agree that . . . there really are not distinct [human] races that one can identify using biological criteria," Graves said.*

---

* Slavery destroyed families. But it also preserved much genealogical and historical data about the enslaved. An historian, Gwendolyn Midlo Hall, PhD, of Rutgers University in New Brunswick, New Jersey, told the conference that because slaves were "property", much more was written about them, in documents, than was recorded about free people at the same times. In the 1980s, Hall discovered that the attics and basements of many courthouses in Louisiana contained documents that were full of information about individual slaves and the slave trade. These depositories had been largely forgotten. Hall visited the courthouses and over a period of many years created a database and a free website: http://www.ibiblio.org/laslave/, in which *laslave* stands for Louisiana slaves. It contains much historical information and biographies, based for example on ads for slaves and dead masters' property inventories. Contemporary blacks have used this resource to trace their lineages back through slavery to Africa.

"*Biological* races are not real," he concluded. "But *socialized* races are as real as a heart attack" [emphasis added]. Or, as powell put it succinctly:

"Race is a scientific fiction and a social fact."

The conference then moved on to the social and personal uses and constructions of "race". The non-scholars present precipitated much of the discussion, which revealed how different individuals used — or ignored — feelings and knowledge of "race" in their sense of self. However, this is too far afield of the history of the ABG to be covered here. The complete transcript of the African Genealogy & Genetics Conference can be found at *http://web.archive.org/web/20070317131828/ http://www.bioethics.umn.edu/afrgen/html/Themythofrace.html*. Five years later, in a state-of-the-art review in the journal *Science*, Troy Duster and a dozen colleagues wrote: "These [genetic] tests should not be seen as determining the race or ethnicity of a test-taker. They cannot pinpoint the place of origin or social affiliation of even one ancestor with exact certainty."

David Zimmerman

23

Finish Line!

MAJOR CHANGES HAD BEEN made in the African Burial Ground (ABG) project in 2002. More were soon to come. Little new research was attempted. The aim rather was to gussy up the findings to date.

Michael Blakey's wish to avoid anonymous peer review of the bioanthropology final report was agreeable to Howard University and the General Services Administration (GSA). A contract modification between GSA and Howard on January 8, 2003, stipulated that the previous, independent peer reviewers would be replaced by a three-person review team. Two members would be chosen by Howard; the third by GSA. All three would have to be "mutually acceptable to both parties," according to *Activity Report* (January 31, 2003).

"The review board will conduct an in-depth review of the scientific merit of the report, determine its adherence to the scope of work, and provide a quality assurance review, prior

SANKOFA?

to its submission to GSA and the regulatory agencies." The final history and archaeology reports would be similarly vetted by appropriate reviewers.

In February, GSA issued a news release confirming the project's new directions, and promising that the bodies would be reburied, ceremoniously, within the year. GSA had "hired" the Army Corps of Engineers (ACE) as technical advisors, according to the February 20, 2003, news release. "GSA has retained the expert services of ACE to provide technical assistance and *day-to-day project management* of the ABG Project," the news release said [emphasis added].

In its announcement, GSA said:

"[W]e are applying renewed leadership and priority to the ABG project, and the results we are achieving will lead us to [its] successful conclusion," a high GSA official is quoted as saying. "We are definitely focused on completing the ABG project prudently, expeditiously, and with dignity."

Another official added:

"We have been working diligently to re-engage the parties who are working together to complete this project successfully .... GSA is now poised, *for the first time in a decade* to complete the scientific work . . . which is long overdue." [emphasis added].

Much of the progress resulted from a change at the top of GSA in Washington. In May 2001, President George W. Bush appointed a new GSA administrator, named Stephen A. Perry. He was a black man, a businessman from Ohio. GSA public relations woman Renée Miscione said later, "he took a direct and personal approach to resolving the issues and concerns related to this project, including making himself available to the Descendant Community to address their specific concerns . . . "

Added GSA official Peter Sneed:

"He was the first guy from Washington who got everyone talking to each other. He never faltered. He held Blakey's and Howard's feet to the fire.... He rolled his sleeves up and cut through the B.S."

Since he no longer was employed by Howard, Blakey was being paid directly as a private consultant by GSA's New York regional office, at a rate of $480 per diem. When work recommenced, he was paid, at a slightly higher rate, as a Howard consultant.

His successor, ACE anthropologist Michael Trimble, PhD, refused this writer's request for an update interview. A GSA official, John Scorcia, said: "We're not interested in doing that right now!" The GSA's penchant for secrecy clearly had not changed.

The deadline for the skeletal report had been moved forward again, to June 2, 2003.

The DNA and bone chemistry studies had now been totally separated from the ABG bio-anthropology research. The DNA research was picked up by the ABG Project's parasitologist-*cum*-molecular-biologist Fatimah L.C. Jackson, PhD, at the University of Maryland in College Park, just outside Washington, D.C. She had earned her doctoral degree at UMass Amherst.

"We're knee deep in DNA extraction, analysis, and interpretation," Jackson said late in 2002. She added that none of her findings would be ready for the bioanthropology final report. Besides, she said the government wasn't paying for her work, which she was supporting with her own grants and private financing.

In an interview in her office early in 2003, Jackson said that she was busy setting up a collaborative DNA laboratory in Cameroon, in West Africa. This would provide a source of

African DNA specimens. It would be the first human DNA bank in Africa, she noted.

The efforts to trace Americans' lineages to Africa are thus far imprecise. "We can't say that the skeleton over there is a Yoruba," Jackson said. "But what we think we can do is to come up with regions" in Africa where an American black person's ancestors may have originated. What is more, she added, these interpretations of DNA similarities will be "probability statements." They will not be identification. "We're thinking about regions" or "*regional clusters*" of genes, she said. DNA alone, without familial or historical records, is not likely to lead one to one's roots, Jackson added. She noted:

Bone sections from the ABG remains are being held out from reburial, so that genetic and bone chemistry studies still can be done. There are 248 DNA/molecular genetics samples and some 2,000 tissue samples available for follow-up studies.

BY AUGUST, BLAKEY, MACK, and their close associates had submitted a draft of their new report to the three-man Skeletal Biology Review Board, which returned it to them with comments. These comments will be incorporated into the report, Activity Report said (July 31).

At the end of September, the remains were returned to New York City with much fanfare, as described in the subsequent *Activity Report* (October 31):

> The Rites of Ancestral Return commenced on September 30, 2003, with a tribute ceremony at . . . Howard University. [F]our individuals representing all of the deceased were then taken to Baltimore, Wilmington, Philadelphia and Newark on October 1st and 2nd for tributes in each community. On Friday, October 3, 2003, the four coffins were taken by flotilla from Jersey City to Wall

Street, where, after a brief ceremony, they joined a procession of five horse-drawn wagons carrying the remaining coffins up Broadway to the memorial site at Duane Street.

The wagons were escorted by members of ... diverse community organizations, who, acting as pallbearers, transferred the coffins from the wagons to the crypts [that had been prepared in the African Burial Ground]. After the coffins were placed into the crypts, a viewing and vigil commenced to allow members of the public an opportunity to pay their respects. On the following day, October 4, 2003, an internationally attended public tribute was held to conclude the Rites of Ancestral Return ceremonies with the lowering of the seven wooden crypts containing the human remains and associated artifacts of 419 individuals.

In an emotionally moving ceremony the remains were returned whence they came: an unbuilt plot adjacent to the 290 Broadway federal building. They were returned to the ABG. Besides the reburied remains and artifacts, the site is estimated by archeologists to contain 200 to 300 intact graves, buried under 25 feet of fill.

Two-and-a-half years later, on February 27, 2006, President George Bush proclaimed the unbuilt plot, less than half an acre, a National Monument. (The larger, seven-acre burial ground that encompasses it had been designated a National Historic Landmark in 1993.) In 2007, a walk-through monument memorializing the ABG population was opened next to 290 Broadway. It cost $5 million, according to the National Park Service. These words are engraved on the monument's wall:

> "For all who were lost,
> For all those who were stolen,

SANKOFA?

For all those who were left behind,
For all those who are not forgotten."

Exclaimed ABG historian Howard Dodson: "This is an important milestone in the whole memorialization process."

The project's GSA boss Alan L. Greenberg later declared: It "consumed GSA for years, taught us all a lesson in sensitivity, and ultimately resulted in a major research project and first class memorial . . .."

In an exultant interview at the Reburial, Blakey acknowledged, inadvertently, what went wrong: "[T]he Sankofa symbol is so perfect[!] It resonates so completely with the ABG. It has to do with the idea that you need to go back and search in the past, to let the past be a guide . . . . That you have to look backwards in order to look forwards . . . . to revere ancestors and respect elders — all these . . . ideas . . . about the relationship between past and present are wrapped up in that symbol."

But science does *not* look backward reverentially to find its guidelines. Trying to act on this contradictory agenda is what stymied Blakey.

The scientific skeletal biology report remained unfinished in 2003. Blakey submitted a second draft to the Skeletal Biology Review Board. He also sent it to the participating agencies for comment: The agencies' comments were sent to Howard. Blakey's team then provided a third draft.

The deadlines for finishing the report were slipping again late in the year. A first draft of Blakey's Final Report manuscript was now due at the end of February 2004. When completed, Blakey again sent it to the Skeletal Biology Review Board and the participating agencies. In June, *Activity Report* confirmed that it was undergoing "technical editing" before its submission as a final manuscript.

By autumn, a Final Report, version 2, had been sent to the Review Board and government agencies for a "final review," *Activity Report* noted.

Early in 2005, Blakey and Howard submitted the Final Report to GSA, which "formally accepted it," according to *Activity Report*. GSA made it available in the Office of Public Education and Interpretation Reading Room at 201 Varick Street in Manhattan and posted it on the ABG website: http://web.archive.org/web/20120721214914/http://www.africanburialground.gov/ABG_FinalReports.htm, from which it can be downloaded. It had been 12 years since GSA approved Blakey's Research Design, and it was 9 years after the date he had originally promised to deliver the Final Report.

THE AUTHORS OF THE ABG History Final Report, Edna G. Medford, PhD, of Howard University and four others, whose work has not been described in these pages, said that Blakey and Lesley Rankin-Hill, PhD "suggest" that violent punishments were meted out by whites — which is an undisputed fact. The historians continued: "In their study . . . [Blakey and Rankin-Hill] found evidence of numerous fractures to both men and women that are suggestive of violent treatment." Besides Burial #25, the woman shot with a musket, historian Medford and her associates cite five other burials:

The remains of Burial #364 "appear to be mutilated," Medford says. In four other burials (#330, 331, 362, and 372), only heads were found, "suggesting, perhaps deliberate decapitation." But none of these possible atrocities was alleged in the burial descriptions in Blakey and Rankin-Hill's Final Report.

Neither the historians nor Blakey and Rankin-Hill per-

SANKOFA?

formed scientific analyses of the five sets of remains in question. Cassandra Hill analyzed all but #330, she said in 2006:

Re: Burial #331. "There's hardly anything left," she said, studying the photo in the skeletal biology report. "But if there's just a skull, so what! . . . Finding just skulls never means anything [about decapitation] without other forensic evidence. Are there cut marks around the cranium [skull]? Are there cut marks on the cervical vertebrae [neck bones]? The rest of the remains may just be gone" — dissolved into the earth.

Re: Burial #362. "Photo shows that a second grave shaft had been dug partly into #362, and its skeletal remains may have been unearthed and discarded at that time," Hill said. "There's no evidence of mutilation in this one!" The same can be said for Burial #372.

"The unmitigated presumptuousness of this just astounds me!" Hill declared. "It stupefies me!"

In short, according to Hill, discovery of a skull without the skeleton is not evidence of a beheading.

Later in a phone interview in 2009, Medford acknowledged, "We found no definite evidence that any" of the dead "were treated violently." She added: "If I did say that [in the History Final Report] then I was wrong!"

What do the archeological researchers — as distinct from Blakey and his anthropological team and the historians — conclude from their study of the burial ground? They claimed that the "conformity" among burial practices that they documented "speak[s] most importantly of the individual's relationship to others — to family but also to a larger community." They added:

"We think the cemetery provided a way for a community to form, through communal performance of a fundamental rite of passage. If, via the archeological record, we are seeing mainly the shared aspects of mortuary behavior then we have a remarkable window on a critical historical process.... Since the African Burial Ground would have been one of the few sites where black men, women, and children could act communally on each other's behalf, it would have been a key place and institution for the continual incorporation of diverse newcomers into the fold."

As stipulated in contract amendments between GSA and Howard, some work remained to be done: the university was tasked with preparing a single volume "integrated report," based on the three final reports. Also, Howard was to prepare a "general/public audience publication" on the project.

SANKOFA?

24

WHAT THEY FOUND

DID THE RESEARCH ANSWER the questions that Michael Blakey posed at the start? **The first question was: Where did the African Burial Ground (ABG) people come from?**

It has not been answered.

Blakey promised DNA analyses and comparison of the findings with contemporary African populations. But the DNA studies for the most part weren't done. Only 48 sets of remains were examined in this way, by Matthew George, PhD and Rick Kittles, PhD, in a pilot study. Fewer than half yielded usable DNA that could be compared to reference samples. They showed that some of these skeletons belonged to one of three large territorial groups that include most of Africa and parts of Europe and Asia as well. A few could be further regionalized to southwest Africa (where many slaves are known to have originated). But the localization of ancestral populations to specific regions or groups like the Kakongo or the Nygoyo simply was not done.

Question two: What was the physical quality of the enslaved population's lives?

For most, the ABG project data suggest, it was not very good. Or, worse. It appears to have been worse than the quality life of white remains disinterred from the nearby Trinity Church Cemetery. Whether it was better or worse than that of Native Americans of that era or of Africans who remained — free — in Africa is unknown.

If Blakey had adopted the methods developed by Jerry Rose, PhD, and his colleagues, these comparative questions, and many others like them, could have been answered objectively (see Chapter 9). Placing the ABG population into context with other modern and ancient burials would have been a noteworthy scientific advance — and perhaps of keen interest to the Descendant Community and others.

The third question was: What biological and cultural transformations took place as Africans adjusted to life in New York?

The Final Report provides very few answers. To trace changes through time and generations, Blakey and company would have had to analyze the remains in chronological order. This they didn't do. So there is no way to know what biological changes, if any, took place as the ABG population adjusted to life in America.

Neither are there many data to answer Blakey's fourth major question:

How did these people resist slavery?

He sought, but did not find, evidence of executions, brutality, and imprisonment in the skeletal remains. The one skeleton, a woman, that clearly shows violent death, due to a musket ball in the ribs (Burial #25), does not reveal whether she was killed by a white person or a black — by a master, a husband, a lover, or someone else.

All biological judgments about the health, longevity, and well-being of Manhattan's 18th-century black population require researchers to use the data they obtained by studying the ABG remains in order to infer what happened in the wider community to people whose remains have not been recovered — and may never be. If, for example, 90 percent of the recovered skulls had bad teeth — dental cavities and broken teeth — then one might reasonably infer that most black people living in Manhattan in that period had bad teeth. But if 9 out of 300 skeletal remains had suffered a broken leg — also a detectable injury — can one infer that 3 percent of the overall population had broken a leg?

This kind of analysis is called paleo-demography, the vital statistics of old or ancient (paleo) populations. The discipline is used, for example, to determine the sex ratio in a population; the proportion of men, women, and children; the distribution of ages at death; average life span; and so on.

"[T]he quantity and quality of data available for this study is [sic] sufficient for an accurate reconstruction of the larger, living African community of colonial New York City," declared Lesley Rankin-Hill, PhD, the first author of a chapter on paleo-demography, deep in the Final Report, and her co-authors, including Blakey.

But then, she and her colleagues explored why these data may not be "sufficient". (Paleo-demography is one of two sources for such data; the second is historical documents, such as churches' burial records, albeit there were no such records for the ABG.) Paleo-demography has been subjected to several rounds of "intense criticism," Rankin-Hill wrote. These critiques have been followed by "discourse and proposed solutions" for the discipline's "intrinsic problems."

One of these critiques, in the 1980s, charged that paleodemography was fatally flawed — and should be abandoned — because the criteria for skeletal aging reflect the aging of known reference populations and may not apply to ages of anonymous bones in a particular cemetery. Also, age estimates of adult remains "lack sufficient accuracy to allow for demographic analysis."* A cemetery's unknown mix of individuals with a variety of susceptibilities to illness and death makes it "almost impossible to interpret . . . aggregate data."

"Notwithstanding the limitations," Rankin-Hill and her associates wrote, "with clear knowledge of the limited 'value' of life table analysis, some basic observations will be presented here." This is hardly a strong endorsement for the project scientists' findings on the ABG population. What they suggest is that survivorship and life-expectancy curves are similar to those constructed for the Cedar Grove Cemetery and other

---

* This is the researchers' summary of the ABG paleo demographic studies:
- Mortality was highest for:
  infants 0-5 months (9.6%).
  adults 30-34 year olds (9.1% annually).
  adults 45-49 year olds (8.3% annually).
- Young adults aged 15-19 comprised 8.8% of the sample.
- A differential mortality trend by sex was observed:
  62% of the females died by the end of the fourth decade.
  45% of the males died by the end of the fourth decade.
  Female mortality (37.6%) peaked at age 30-39.
  Male mortality (34.3%) peaked at age 40-49.
- Subadult mortality was 43.2%.
  39.2% died during the first year of life.
  16.2% died in the second year.
  55.3% of all subadults died by age two.

old, black cemeteries.

Finally, does the Final Report answer the many detailed and specific questions Blakey had promised the Descendant Community and others that he would research?

Below are a few of those promises, and the answers, based on the Final Report:

A section of the Final Report, dramatically titled "Life and Death in Colonial New York" (300 pages) covers much the same material as the other sections. But this time around, there is much less verbiage and many more data, in great detail. First is a report on children's health and dental development; Blakey and dental expert Mark Mack are the lead authors.

The major premise is that defects in the formation of tooth enamel (as described in Chapters 11 and 12 above), called hypoplasias, or areas of disturbed growth, are the manifestation of "metabolic disturbances of malnutrition and disease elsewhere in the body." They are "general stress indicators," and they are particularly valuable for anthropologists because this evidence endures as long as the teeth — the body's hardest structures — persist. Because dental enamel is not made of cells, it is not erased by the cellular renewal processes that affect the rest of the body. What is more, hypoplasias can be readily compared to teeth from other skeletal collections.

"Hypoplasia [thus] provides an estimation of stress severity and/or duration by the size of the malformation," the researchers wrote. "With rare exception, dental enamel hypoplasia is a result of systemic metabolic stress associated with infectious disease, insufficient calcium, protein, or carbohydrates, and low birth weight," which are "characterized together as 'general stress.'"

The ABG sample (99 individuals) had a high level of hypoplasia, as do remains from other historic black cemeteries.

Many data and many comparisons have thus been engendered. The conclusion of the chapter on nutrition and bone infection, similar to other chapters in the report, says, blandly, "the information presented here suggests that infectious disease, in conjunction with inadequate nutrition, was another source of chronic stress for the enslaved population of the ABG . . . [S]tudies of disrupted growth and development and early mortality are consistent with these findings."

A long chapter on skeletal manifestations of forced labor ends up similarly blandly: The "most consistent results," Blakey and three other scientists reported, are that "strenuous labor started at an early age" for at least some ABG individuals. Arthritis of the legs and ankles "suggests" high general stress, such as walking on rough ground or inclines carrying heavy loads. But none of these changes is specific and clearcut. So, "linking individuals with specific occupations would be imprudent," the researchers wrote, "when one considers the wide range of possible activities that might affect a single marker" of bone degeneration. What is more, individuals are built differently, and they vary in the way they work.

The penultimate planned area in the Research Design was chemical studies of teeth, and possibly also bone, for trace elements that could indicate where a person was born and raised. For want of money, only preliminary studies were conducted on samples drilled from teeth from 40 ABG individuals; most of the teeth were molars.

One of the other, most clear-cut findings involves the element lead. It often is found as an impurity of tin, one of the metals used to make pewter. Dishes and drinking vessels in colonial New York were often made of pewter. But lead is

SANKOFA?

rarely found in comparable African tableware.

When Mack and his colleagues assayed lead levels in samples from the 40 teeth, they found that people with modified teeth, presumably born and raised in Africa, had far lower lead levels than teeth from people with nonmodified teeth, who are more likely to have been born and raised in the Americas. The average level of lead in the nonmodified was six times higher than in the modified teeth.

Osteologist Susan K. Goode-Null, PhD, analyzed at length the skeletal growth and development data as indicators of ABG individuals' general health. But she concluded that skeletal growth "does not provide a clear picture of cause effect [sic] in relation to growth status." Nevertheless, she (and her co-authors) wrote, her analysis "does allow a few general conclusions":

- The measurable individual heights "clearly indicate a population that was not reaching its growth potential." She then reported, hedgingly: "*Given that* growth status is *often used* as a proxy for overall population health, *it is not injudicious to put forth* that the overall health status of the ABG population was poor" [emphases added for hedges].
- Evidence of biomechanical stress in children as young as four "indicates that children . . . were engaged as laborers."
- Physical development "was affected negatively" by its [sic] social milieu.

THAT IS ESSENTIALLY WHAT is covered in the Final Report. In it, Blakey and Rankin-Hill were not very forthcoming about how they prepared the massive report. But given that Blakey is listed as an author on 13 of the 14 sections, one can surmise that he did much of the writing, while she did much of the editing.

The report includes a slap at Blakey's professional colleagues for the "racist and inhumane anthropological practices of the past." And, a brief, final discussion, for which Blakey is first author, characterizes the report's findings as "shadowy evidence" — which seems correct.

The report has a 60-page bibliography: some 600 references to the authors' and contributors' own research and others'. Finally, there are 146 pages of photographs of each and every set of skeletal remains, taken before exhumation in Manhattan or on the examining tables at Howard. A handful of documents completes the report.

How does the ABG population's health compare to that of other ancient and historic cemetery populations in the Americas? Unfortunately, Blakey's failure to make scientific assessments of the remains, and the content and format of his findings render the ABG data incomparable to the data in Rose and Steckel's Mark I or later repositories (described in Chapter 9), anthropologist Rose said later.

The Final Report's lack of relevance was publicly displayed late in 2005, when the New York Historical Society in Manhattan mounted an exhibit "Slavery in New York," inspired in large part by the ABG discoveries. The human remains had since been reburied, so the exhibit lacked the gripping effect reported by black leaders and others who had glimpsed them before exhumation or during the decade of scientific study.

The exhibit appears to have contained little or no new information from the Final Report. But the exhibit curator, historian Richard Rabinowitz, PhD, said in a press conference in mid-November, "we were able to use research material to discuss [in the exhibit] nutritional deficiency," bone deformities due to heavy labor, and other such findings. The bones were "enormously revealing," he said. But: None of this information is attributed to the Final Report, and most, if not all of it, could have been gleaned by simply eyeballing the remains or reading about them in press releases reporting their discovery and disinterment.

Blakey had described some of this information in an interview published in *USA Today* on September 15, 1992. This was one full year before the remains were removed from the Bronx and sent to Howard for study! So, it appears that the research at Howard University contributed little or nothing to the exhibit. Neither did the exhibit contain any artifacts from the ABG, according to Historical Society publicist Laura Washington. They had been reburied.

The exhibit, like early pronouncements on the ABG project, exaggerates the conditions of New York slavery — and in much the same language. Exhibit literature describes New York as "the capital of American slavery" and argues that "only Charleston, South Carolina rivaled New York City in the extent to which slavery penetrated everyday life." It claims, too, that New York slavery was as harsh as South Carolina's.

A *New York Times* writer, Edward Rothstein, pointed out, however, that slavery was so different in the two places "that the comparison doesn't begin to capture the differences." He added: "The exhibit's argument about New York City's centrality to slavery . . . can seem overstated . . . ." One key difference: In the South, scores of slaves tended huge industrial plantations; in New York, white families tended to have one or two slaves at most, who did household chores and ran errands — and so may not have been economically highly productive, as the exhibit, like the ABG literature, claims.

DAVID ZIMMERMAN

25

IS IT SCIENCE?

THE "SKELETAL BIOLOGY FINAL REPORT" is massive: 830 pages. And it is filled with data, many of which well may be valid. The question is: Is it science?

It is a matter of procedure and protocol that determines whether colleagues — fellow scientists — will accept a work into their realm and their discourse, whether or not they agree with its conclusions. Is the Final Report reliable? Can they trust it and use it, or try to confirm or refute it with their own research? Or is it not? — in which case, it can be ignored, because it's not science, even if it may yet prove to be true.

Publication is the step that transforms research into science. The science publisher, or, usually, his or her editors, are the gate-keepers.

Most science is presented to these legitimators in the form of reports — papers — that describe one or a short series of related experimental discoveries, or findings. These typically are brief, very specific communications; for example, James Watson and Francis Crick's identification of the double-helical structure of DNA — one of the 20th century's most important discoveries — was published in 842 words in the journal *Nature*, in London.

The most prestigious scientific publication in the United States is the journal *Science*, published by the American Association for the Advancement of Science in Washington, D.C. The longest research reports *Science* will accept are up to about 4500 words, and they "are expected to present a major [scientific] advance." This is about 20 typed, double-spaced pages, compared to the many hundreds in Michael Blakey's Final Report. The report's primary producers, Blakey himself and Lesley Rankin-Hill, PhD, are listed — as the top editors — but editors are not authors. This alone might disqualify the Final Report as science.

The report's individual chapters are headed by the names of one or more authors; most often Blakey is listed first. The chapters could have been — perhaps should have been — submitted individually for scientific publication in one or more journals (assuming they could be reduced to manageable numbers of words). In the normal course of events in science, that's how multi-authored volumes, like the Final Report, are constructed. They are made up of short papers or reports that already have been validated by publication elsewhere as journal reports. Some chapter authors may present wider amounts of information than can be contained in a research article. But the key elements in these chapters already have been validated as research reports, which then are cited in the longer article.

Blakey and his associates, however, largely failed to publish research articles along the way. Based on the Final Report's bibliography, for example, Blakey has at most one such original research paper, written with a colleague, which was published in 1994, when the African Burial Ground (ABG) studies were barely under way. Rankin-Hill has no such original publications listed in the bibliography. Mark Mack has one publication, a one-page abstract in collaboration with Blakey, and also an oral report with Cassandra Hill and Blakey that is validated as science because Hill presented it at an annual meeting of the American Association of Physical Anthropologists, which published the abstract. Rick Kittles, PhD, has one abstract with Blakey and others. And so it goes: Very, very few original scientific publications, partly because Howard University, at the start, required all such individual papers to be approved by Blakey — and he apparently didn't want anything published ahead of the Final Report.

The ABG Physical Anthropological Peer Review Panel's 1993 imperative, "The ABG data...needs to be made available to the professional community *as work progresses, and prior to reburial and publication of a final technical report*" [emphasis added], thus was not met.* Despite huge expenditures of public funds, very few of the ABG findings reached other scientists or the public.

Individual researchers like Mack and Hill may, of course, be able to publish their specific findings, now that the Final Report has been published. (Mack died in an auto accident in May 2012.) But most journals won't publish findings that already have been broached elsewhere. The editors of *Science* stipulate, for example:

"*Science* will not consider any paper or component of a paper that has been published or is under consideration for publication elsewhere.... In addition, reporting the main findings of a paper in the mass media may compromise the novelty of the work and thus its appropriateness for *Science*."

The Final Report is characterized by editor Rankin-Hill as

---

* See p. 102

SANKOFA?

a "contract report" [in fulfillment of the contract between the General Services Administration (GSA) and Howard University]. One important arbiter of whether a publication is science is the Science Citation Index® (SCI®), which is published online and in other formats by the Thomson Scientific company in Philadelphia, Pennsylvania. The SCI provides content listings and abstracts from some 5,800 journals — for all intents and purposes, all of science.

Asked if the Final Report, as a contract report, fits Thomson's criteria for inclusion in its SCI, the company's manager of corporate communications, Rodney Yancy, said:

"We cover journals, books, and conference proceedings that contain original research or reviews of the literature of original research. We would not consider this report to be science. It seems to be some kind of listing of research done under government contract. Therefore we would not cover [it]."

In short, by the SCI criteria, the Final Report is not science.

Blakey himself acknowledged the Final Report's lack of scientific standing: In a 2004 update of his Curriculum Vitae, he lists his "refereed" — which is to say scientific — publications separately from the various "reports" he had written. The Final Report is in the latter, but not in the former list. What is more, of his several dozen "refereed" publications, none appears to deal with the technical findings of the ABG project.

Howard University, with financial support from the GSA, of course, might publish the Final Report, in some other format, but, beyond the demerits given above, it appears that it has not yet been reviewed by independent and anonymous reviewers. On the contrary, it was reviewed by known colleagues chosen by Howard and GSA. This is quite the opposite, say, of *Science*'s procedures, in which "papers are reviewed in depth by two or more outside referees . . . . the identities of reviewers are not released."

The more basic problem is not length, nor the report's usually insipid prose. Its stated purpose, to confirm an historical image — of enslaved blacks' brutalized lives in colonial New York — is simply beyond the realm of a scientific study. Specific facts may be discovered or confirmed: Bone and joint wear, for example, show that people worked extremely hard — but they do not show whether this was forced labor or the result of hard work in one's home and garden.

Historical documents show that Manhattan blacks were treated poorly in the 18th century. Bone studies confirm this bad treatment, or at best, blacks' poor living conditions. So Blakey and his colleagues conclude that these people were treated badly — which is where the researchers started. The process is circular, and the new evidence is unsurprising, which is why the conclusions are weak — and dull — as in the following:

- With regard to childhood health and dental development: "These findings indicate that the quality of life for Africans was greatly compromised upon entry into the New York environment . . ."
- With regard to dental disease, diet, and nutrition: "These results provide additional evidence of poor dietary regimens, unhealthy living conditions, and lack of dental care that characterizes the quality of life for the majority of those who lived in bondage."
- With regard to skeletal indications of infectious disease and poor diet: "The information presented here sug-

SANKOFA?

gests that infectious disease . . . was another source of chronic stress for the population of the ABG."

• Etc.

Does the above constitute $5 million's worth of research findings?

The lack of anything new to say regarding science may, of course, have some value for Blakey: It refocuses readers' and others' attention on the nonscientific accusations that he and his associates have made about white people, the United States government, and their "racist" colleagues in anthropology.

This writer is not a scientist and so has no vote on whether the Final Report is science. Time will tell whether the ABG results are picked up and used by Blakey's colleagues — and if so, how they are used. Some may demur. GSA consultant Jerry Rose, PhD, was asked whether, if it were his decision, he would incorporate Blakey's data into comparative health studies of ancient and modern people, like the landmark Mark I study described in Chapter 9, above.

Rose answered, succinctly: No.

David Zimmerman

26

WHAT WENT WRONG?

IN RETROSPECT, MUCH WENT wrong with the African Burial Ground (ABG) research project. Much less was learned, and much more time and money were spent, than anyone predicted at the start. The value of the findings remains to be seen.

The major problem was Michael Blakey. He ran the project. He ran it off the rails when he submitted a faulty first draft of the anthropology report in 1998, and he then had to stand by and watch while the work was rescued by the General Services Administration (GSA) and the Federal government that he had warred against.

What can be said about the Descendant Community, Blakey's supporters, who were supposed to — but failed to — adequately monitor the project's progress: All bear some responsibility for the debacle that followed. The GSA was cowed by blacks' threats of violence. It was distressed by an unwanted interruption in the Broadway building's construction by a group and for a purpose it knew little about. Once GSA was forced by Congress, the National Historic Preservation Act of 1966 (NHPA), and the agencies that administer them, to accept and grapple with the ABG project, the GSA seems to have taken a lazy, laissez-faire position on the project: Give them the money and let them do their thing — which lasted until things got out of hand. One part-time consultant, Jerry Rose, was insufficient oversight for the project.

The scientific advisors who declared Blakey and Howard University qualified for the ABG contract also are to blame.

The five scientists who wrote and signed the Research Design Review Panel report in 1994, which stipulated that "all scientific research [should] be conducted at Howard University," clearly gilded the lily. They spoke, for example, of Howard's "more than adequate" facilities, which, in fact, turned out to be three largely bare rooms totaling 3,000 square feet.*

The panel said:

"We wholeheartedly support the African American-communities' desire that the scientific research be conducted at Howard University. [It] is the only American university that has a large, well documented African-American osteological collection and a nationally recognized faculty with expertise in the interpretation of African American history and culture."

These experts had made a political decision, not a scientific judgment. They grossly overstated Howard's human and physical resources.

Certainly there were other screw-ups and errors by the 200 or so individuals and the dozen agencies that had a hand in the ABG project. But the prime mover, who took charge of the venture and promised to bring it to fruition, is the individual

---

* The panelists were all doctoral-level anthropologists.

ultimately responsible for its success or failure. He is, of course, Michael Blakey. He was in charge from 1993 to about 2000, when he failed the project — and he should answer for its shortcomings. These are a few of the major problems:

Blakey was a skilled rhetorician — an effective rabble-rouser — but he was not a very good scientist. His planning was confused, and his proposals through multiple drafts of the Research Design were not clear. When, finally, he had the remains in hand, if you believe Cassandra Hill, he didn't know what to do with them! Neither, she says, did he provide hands-on work at the necropsy table.

Blakey treated Hill badly, and fired her, even though she appears to have been the only scientist on staff who could have analyzed the remains and reported the findings properly in the first draft and Final Report. How could Blakey, as project scientific director, not have known this? And if he did know Hill's irreplaceable value, why did he shoot himself in the foot by firing her?

More important, why did he then write a dreadfully inadequate preliminary draft, and then submit it to GSA without having had it vetted in house, at Howard? If he didn't know that it was weak work, he may have been blind to its faults or may have had a poor understanding of science. If he had a writing problem that he and his associates couldn't resolve, why didn't he pick up the phone and hire a science writer or editor to help? — there are dozens of them in the Washington area!

Instead of fresh discovery, based on scientific rules and procedures, Blakey appears from the start to have been more interested in using the data to show up and confirm his non-scientific beliefs about how the ABG population lived and died. Science can answer some very particular questions:

SANKOFA?

Does the DNA evidence confirm that John Doe raped Mary?* But science is ill-used when it is asked to explain — and to justify — broad social theories. As is the case of the Soviet Union's adaptational theory of genetics, or the Nazis' belief in racial purity, the "science" performed to support an ideology or social theory usually turns out to be bad science — or non-science. Blakey's approach to the ABG reflects his ideologically and politically biased beliefs, however progressive and salutary they may have been. These beliefs, shared by many in the Descendant Community, twisted the ABG effort from science to a religious-like belief system. Blakey failed when he was unable to bring this system into conformity with the canons of science. In this, he failed to provide his black "ethical clients," the Descendant Community, or the rest of us with the specific and perhaps less dramatic information that could have reliably informed us of this tragic and little-known facet of American history. That's a shame.

One major question remains unanswered: If the ABG anthropological research was not science, as it is professionally and commonly understood, then what was it? Blakey claimed it was a different, new science — *Vindicationism* — which he acknowledges starts out with a bias: It is pro-black and anti-white. This clearly distinguishes it from science per se, which is, or aims to be, unbiased, so that universally valid truths may emerge. Vindicational science, like Christian Science and Creationist science, is outside the scientific realm.

---

* Or, technically, DNA in semen found on Mary's body matched John Doe's, and the possibility that it came from someone else was less than, say 1 in 10 million.

27

FINDINGS

MICHAEL BLAKEY'S PUBLIC AIMS for and anticipated benefits from the ABG project have been well publicized. But he has been reticent to discuss his own professional and personal goals and the nonscientific roles he played.

Most basically, the project provided paid work. It also paid Blakey in prestige: He was chief of a very important research effort, which also brought him special recognition as a Black Community activist. The first was a career maker; if he succeeded, he might win scientific fame, job offers, and invitations to lead other prestigious research ventures. From the Black Community, he could look forward to a leadership role and the perks it could provide.

But there was more — much more! Black people identified, personally, with their buried forebears; Blakey cultivated and encouraged these emotional ties. No surprise, then, that he, too, became deeply engaged with the people these bones had belonged to. How? By rescuing them spiritually. By redeeming them. By vindicating them against the "Euro-American" white world and the "racist" United States government through which blacks had long been oppressed.

Blakey set out to reveal and recreate the African Burial Ground (ABG) lives, depicting them as the pioneer black settlers in America, who triumphed — The last shall be first! — in death. They played a leading role, albeit still in bondage, in building New York and, hence, the United States. Scientists will attempt "to reconstruct the lives of colonial-era Africans," he said. Elsewhere, he added: "We seek to reverse the false histories that deny the African material contributions to the Western world."

Blakey thus was a Messiah freeing his people. He also was their raconteur — their mythic reality-maker. Speaking at an "Africanicity" ceremony, including Akan [Ghanaian] rites, that was conducted at Howard, Blakey said:

"I had the sense that we were in the process of continuing the creation of our culture. [And] that this was a proper way for African diasporic people to proceed . . . ."

Blakey treated the ABG population as the ancestors, the forebears, of the current black population. This is a metaphoric but by no means a scientific or historical characterization. It is not clear how many contemporary blacks in New York and the United States are direct descendants of the ABG population, but it must be a relatively few. At a religious service during the removal of the bones to Howard University, Blakey declared: "[O]ur job [as scientists] is to sit at the feet of those who were enslaved. Our job is to restore them to who they were: their origins, age, culture, and work, and to restore their identities, which were buried and seemingly disguised from us forever."

This was both a cultural and a personal quest: "[T]he

skeletons that we examine are full of powerful information that makes the lives of my ancestors more real to me."

Blakey added, elsewhere:

"If you can't take care of your ancestors after all this time, what can you do!"

He wrote:

"You have to share the work and the credit. By doing all of those things I feel that we will have a much more sophisticated view of the ABG, which is to say that [t]he people buried there will be known, understood, and will have identities.... There will be a richness, a depth, and a breadth to their identities that is largely a function of the depth, breadth, and sophistication of the research."

Blakey has an assertive, aggressive, and, some colleagues say, an unpleasant personality. From a scientific point of view, no matter: A researcher's personality has no direct relationship to the value of his or her work. Even an ogre can do good science!

But Blakey had difficulty working with colleagues who challenged his viewpoint. This cost him and the project dearly. The record suggests to me that he never adopted, or adapted himself to, the scientific mindset and its *weltanschauung*.

Blakey acknowledges the disjunction between science and the path he chose. In his delusional view, anthropology's racist methods were, in 1996, being replaced by a new, "anti-racist method," of which he, Blakey, was the avatar.

Lacking a scientific frame of mind, and, driven by ego and ambition, he easily fell back on Black Nationalism. Its precepts aren't wrong. They simply do not belong in a research lab — or, if present, must be kept widely separate from the science. Blakey, however, was intent on using the science to identify iconic people and events from the 17th and 18th centuries — the hanged victims of whites' hysterical reaction to

SANKOFA?

the fires in 1741, for example — that might be discovered in the remains. Little if anything that was uncovered fulfilled these aims. This may have heightened Blakey's disappointment and anger.

One of Blakey's most egregious stipulations — a self-serving one, since he was almost the sole qualified candidate — was that only black scientists were qualified to successfully analyze black humans' remains. But he himself contradicted this precept when he hired a white osteologist, Cassandra Hill, to lead the hands-on study of the bones.

As indicated in this history of the ABG project, Blakey and his co-workers made very few discoveries of note in their decade of study: They published few findings in scientific journals — as they had been asked to do. They even failed to answer many of the questions they had tasked themselves with at the start.

Blakey's dissertation committee at UMass Amherst had affirmed that Blakey knew what was — and was not — science when they "approved [his thesis] as to style and content."* So it appears to me that either he consciously ignored or suppressed this understanding of science while working on the burial ground remains, or the thesis committee members were undiscerning when they certified his ability. Or both.

If the dissertation committee and the university were incorrect — and Blakey was not qualified for the doctoral degree they awarded him — then they share the responsibility for his later failures. They had awarded him an unwarranted Seal of Approval!

Of course, if the politicians and community leaders who later backed Blakey for the leadership job had listened to those

---

* See p. 82

of his colleagues who questioned his plans, the mistakes might not have been made — and the Descendant Community and all the rest of us might now know much more than we ever will about the lives of the 400 enslaved New Yorkers.

It's not that the Steering Committee and the GSA and the community behind them hadn't been warned. They had. In one of these alerts, US Army researcher Madeleine J. Hinkes, PhD, had written:

"My first impression is that the project was designed with more of an eye towards 'political correctness' than to good scientific procedure."*

Blakey released very little meaningful information in the first five or six years. The project's Office of Public Education and Interpretation (OPEI) facilitated this stone-walling. Its chief, archeologist Sherrill D. Wilson, PhD, was in touch with Blakey during this period. But she never told readers in her newsletter, *Update*, that Blakey's first draft had been rejected, and had thus thrown the whole project into turmoil. Finally, GSA, perhaps for reasons of its own, blocked or discouraged interaction between Blakey and the media, thereby perpetuating the problems.

The media cooperated by looking the other way. Thousands of news stories, features, and TV reports appeared with OPEI's skillful help. Not one was critical. Not until early 2001, when the problems became completely overwhelming, did the media — meaning mostly the *New York Daily News* and *Newsday* — look in. But they rarely did so again.

The *New York Times* editors have never run a critical report on the African Burial Ground project. (We offered; they declined.) Why not? You should ask them.

---

* See p. 71

SANKOFA?

EPILOGUE

IN THE SPRING OF 2011, the General Services Administration (GSA) released the final, final documents of the African Burial Ground (ABG) project, which was heading to closure. There had been some striking late developments.

Most noteworthy perhaps, Michael Blakey's interpretation of the tack marks on Burial #101's coffin lid had been disproved. Blakey had said, remember, that the tacks formed a West African designed called a Sankofa. He also had claimed, based on little evidence, that this design showed two birds face-to-face, which he interpreted to mean *We look to the past in order to understand the present and build the future.* This design was important he said because it was the most "ethno-linguistically specific . . . evidence recovered from the site."

Others had suggested that the tack marks indicated a date, 1769, perhaps the year of the interee's death, and a heart, possibly an "I love you!" from grieving kin.

The case against interpreting the design as a Sankofa has been made most persuasively, by historian Erik R. Seeman, PhD, at the State University of New York at Buffalo. Seeman studies what he calls *deathways.* In his recent book, *Death in the New World* (University of Pennsylvania), Seeman compared the coffin top design of Burial #101 to the Sankofa, that is stamped on mourning clothes called adinkra cloth, that are worn by Aken people in Ghana.

Equality. Heart-shaped design from coffin lid of Burial #101 (bottom) is the same as design on lid from white burial (top), in Connecticut. So design is not African, and it is absolutely not a Sankofa, as has been very widely alleged.

SANKOFA?

Seeman cites the following problems with Blakey's view that the coffin lid design is a Sankofa:

- There is no evidence that the Sankofa was used in Ghana's 18th-century funerals; adinkra cloths appear not to have been used by the Akens until the 19th century, and, what is more, cloths from that century do not show the design. Its first confirmed appearance was in 1927.
- A second objection: Masters commonly supplied coffins for their slaves, and Burial #101's master, if he had one, would have been unlikely to pay extra for an African design.
- Third, and most important, hearts made of tacks were also a feature on white Americans' coffins in the 18th century. Seeman published a photo of one such coffin lid, with the deceased white man's initials, age, and date of death tacked onto it; he was a distinguished colonial leader in Connecticut. This design, Seeman said, is "uncannily like the design on the lid of Burial #101's."

More broadly, in opposition to Blakey's Afrocentric interpretation of the ABG burials, Seeman writes that most of the 400 individuals "were buried in ways identical to their white neighbors. . . . [I]t is striking," he adds, "that so few of the burials contain African elements."

The ABG historians, agree with Seeman: "[T]he deceased were laid to rest in a manner not unlike that of white New Yorkers."

ARE THE FINAL REPORTS — the product of such great cost and effort — now in the public realm, as Blakey, Howard University, and GSA had promised? Yes, they are. Howard was also on the hook for an "integrated" report, The Skeletal Biology, Archeology, and History of the African Burial Ground: A Synthesis of Volumes 1, 2 and 3, and also a shorter "general audience report," The New York African Burial Ground: Unearthing the African Presence in Colonial New York. GSA would pay for this work.

Instead of writing these reports in-house, Howard farmed them out to a cultural resource management firm, Statistical Research, Inc., in Tucson, Arizona. It, in turn, assigned the writing to its research director, archeologist Michael Heilen, Ph D. At the time, he was working on a project comparable to the ABG: The disinterment and study of 1,100 intact burials – Hispanics, Native Americans, and Anglos – from a 19th-century cemetery in Tucson.

Noting that the ABG was a "complicated" and "politically charged" venture, Heilen said, Howard "restricted us from talking to the ABG Team. I was never able to speak directly to Blakey at all," he explained by phone.

The most important findings, in Heilen's view, were Cassandra Hill's recording of the physical conditions of each of the 400 sets of ABG remains, for which he said there are no comparable data from other sources. Asked if, overall, the ABG reports that he had labored to synthesize were good science, Heilen answered: No.

The integrated report that Heilen wrote was set in type and posted on the ABG website. The General Audience Report also was posted there.

These final reports on the ABG Project were, editor Heilen said, "derived primarily" from the work by Blakey and his colleagues. But Heilen added new material, as well as his own opinions. So what has emerged is one archeologist's para-

SANKOFA?

phrase of Blakey's mammoth research work. (Blakey, he said, has declined to have it published under both of their names.)

Heilen has changed the project's focus in his version. For Blakey and the news-reading public, it has always been an anthropological venture, based primarily on the disinterred bones. But Heilen, taking into account the archeological and historical reports, said it was "primarily an archeological project" — sticks and stones rather than teeth and bones — and he has edited it according to archeological standards. Hence, there are detailed data on the contents of whore house privies that succeeded the burials, as the cemetery was forgotten and the land turned to other uses in the 19th and 20th centuries.

"The research ... was ground-breaking in its intent and content," Heilen wrote, setting the tone for his synthesis of the findings. He cites Blakey's 1992 testimony to representative Gus Savage's subcommittee that the (then unacceptable) research design would lead to "the most sophisticated and comprehensive skeletal biological project ever conducted" (see pg. 49). Has it?

So, WHAT DID IT cost in the end? GSA's contract with Howard University closed out at $8,161,623.30. The total cost, according to one ex-GSA official who asked not to be named was $40 to $50 million. More recently, the project's long-time boss, GSA Assistant Regional Administrator Alan L. Greenberg, now retired, said, "in my estimate the ABG cost the government $100 million in mitigation and memorialization expenses .... The costs ... far exceed any GSA news release. Take it from me, I signed a lot of the checks." Greenberg said in a phone interview that only a small part of this money went into the memorial. But, assuming that the $100 million was divided evenly between mitigation and memorial, that means that the mitigation — work on the remains — cost $125,000 for each set of bones ($50,000,000 ÷ 400).

Michael Blakey's career has blossomed, according to Google. He is now (2013) a National Endowment for the Humanities Professor of Anthropology, and chief of William & Mary's Institute for Historical Biology, at an annual salary upwards of $150,000.

Cassandra Hill has fared poorly. She remains in Tuscaloosa, Alabama, where the economic recession has curtailed her research work.

One final note: Now that the ABG project is finished, Michael Blakey no longer controls the data. Late in 2011, Cassandra Hill dug out of storage her 15-year-old notes on the remains, and reviewed them. I hope she decides to edit them for publication, so that information gleaned from these bones of enslaved early Americans may be permanently preserved—and so be honored in the annals of science.

SANKOFA?

About The Author

David Zimmerman grew up in the Midwest. He attended the University of Chicago and Brandeis University, where he received his bachelor's degree in History and English. Zimmerman worked for newspapers and magazines in New York City for many years. He wrote a half dozen books. Now he works and resides in Vermont.

SANKOFA?

SOURCES

These abbreviations are used here:

ANTHROPR = African Burial Ground Project [Blakey, Michael L. et al.], *Skeletal Biology Report, First Draft*. New York: U.S. General Services Administration, Region 2, August 5, 1998, 3 vols. http://web.archive.org/web/20120721214914/http://www.africanburialground.gov/ABG_FinalReports.htm

ANTHROFR = Blakey, Michael L. and Lesley M. Rankin-Hill, eds., *Skeletal Biology Final Report*, New York, U.S. General Services Administration, Northeastern and Caribbean Region, December 2004, 3 vols., http://web.archive.org/web/ 20120721214914/http://www.africanburialground.gov/ABG_FinalReports.htm

ARCHEFR = Perry, Warren, Jean Howson, and Barbara Bianco, eds., *Archeology Final Report*. New York: General Services Administration, Region 2, February 2006. http://web.archive.org/web/20120721214914/http://www.africanburialground.gov/ABG_FinalReports.htm

HISTFR = Medford, Edna Greene, *History Final Report*. New York: General Services Administration, Region 2, Nov., 2004. http://web.archive.org/web/20120721214914/http://www.africanburialground.gov/ABG_FinalReports.htm

EPIGRAPH

p. 10, p. 00: "Negroes must be ... creative reconstruction." King, Martin Luther, *Stride Toward Freedom*. (New York: Harper & Row, 1968), p. 223.

"We must stop ... of science." Gore, Al, *The Assault on Reason* (New York, Penguin, 2007).

INTRODUCTION

p. 8: "We consider this sacred ground," *New York News*, May 15, 1992,

p. 15.

p. 8fn: Cobb was a .... of crime.   Barbour, Warren T., Musings, on a dream deferred. *Federal Archeology Report*, Spring, 1994, p. 12.

GRAVE FINDINGS – Chapter 1

p. 17: In the late ... potter's field. Cantwell, Anne-Marie and Diana DiZerega Wall, *Unearthing Gotham: The Archeology of New York City*. New Haven: Yale, 2001, p. 279.

p. 18: By century's end ... the population. Berlin, Ira, and Leslie M. Harris (eds.), *Slavery in New York*. New York: The New Press, 2005, p. 63.

p. 18fn: The first resident ... from hispaniola. *Ibid.*, p. 24.

p. 18fn: "Beyond the commons .... and outcries." cited in Cantwell and Wall, op. cit., p. 279.

p. 19: A less dire .... their midst. HISTFR, p. 51.

p. 19: Englishmen from Massachusetts ... now Connecticut. Shorto, Russell, *The Island at the Center of the World*. New York: Vintage, 2005, p. 260.

p. 19: Because of the ... poorly recorded. ARCHEFR, pp. 133-135.

p. 19fn: The adjacent property .... Franklin Roosevelt. *New York Times*, July 24, 2011.

p. 23: But all the ... parking lot. ANTHROFR, p. 4.

p. 25: The researchers knew ... our time." Hansen, Joyce, and Gary McGowan, *Breaking Ground, Breaking Silence*. New York: Holt, 1998, p. 6.

p. 25: "Two centuries ago .... its boundaries!" David Dinkins, cited in Cantwell and Wall, op. cit., p. 282.

p. 26: "It's bad enough ... in death." David Paterson, quoted in Cantwell and Wall, op. cit., p. 283.

p. 27: The GSA quickly .... of removal. *New York Times*, October 9, 1991.

p. 27: The agency's regional ... this project." *New York Times*, December 6, 1991, late edition.

PROCEDURES – Chapter 2

p. 28: The program is .... American people. National Historic Preservation Act, Section 1.

p. 29: At the national ... in point. *Introduction to Section 106 Review*. Reno, Nevada: Advisory Council on Historical Preservation and University of Nevada, January 2003, especially Tab 6.

p. 30: The New York .... new building. Advisory Council on Historical Preservation, *Report to the President and Congress of the U.S.*, 1992. p. 65.

p. 32: The search for .... resource management. Heilen, Michael, ed, *The Skeletal Biology, Archeology, and History of the New York African Burial Ground: A Synthesis of Volumes 1, 2, and 3*. Washington: Howard University, 2009 p. 273.

p. 33: "In it, GSA . ... Burial Ground." Advisory Counsel, op. cit.

p. 34: "The excavation of .... the ship." Cantwell and Wall, op. cit. p. 234.

p. 36: It was the .... in 1977. American Broadcasting Company display in exhibit "Slavery in New York," New York Historical Society, October 7, 2005 to March 5, 2006.

p. 36fn: Malcolm X had ... automatically dead." Marable, Manning, *Malcolm X*, New York: Viking 2011, p. 277.

p. 37: [T]he remains ... original cemetery .... General Services Administration, Region 2, "Research Design in Archeological, and Bioanthropological Investigation of the African Burial Ground," April 22, 1993, pp. 23-34, sec. 2, 4, 1.

p. 38: "driving a car ... or destination." Cantwell and Wall, op. cit. p. 285.

p. 38: "The fact that ... Turkel exclaimed." Cook, Karen, *Village Voice*, May 4, 1993 pp. 23-27. This is the most complete journalistic report to appear during the ABG project's early stages.

A DEFINING MOMENT — Chapter 3

p. 40: The first indication ... Washington, D.C. African Burial Ground-Project, "Project Chronology and Timeline," off-set n.d., p.1.

p. 41: "According to a ... Burial Ground," *Ibid*.

p. 41: The ABG population .... African-American community." Howard University and John Milner Associates, "Revised Draft Research Design for Archeological, Historical, and Bioanthropological Investigations of the African Burial Ground, prepared for General Services Administration Region 2, New York, April 22, 1993. p. 2. Courtesy of the Schomburg Center for Research in Black Culture.

p. 42: Michael Louis Blakey .... an anthropologist. "Michael Blakey," in *Current Biography Yearbook*, 2000, pp. 60-63.

p. 44: In the thesis .... us both." Blakey, Michael L., *Stress, Social Inequality, and Cultural Change: An Anthropological Approach to Human*

Psychophysiology. PhD diss., University of Massachusetts, Amherst, 1985. pp. vi-vii.

p. 45: "a left-leaning black ... student years." Cook, Karen, *Village Voice*, May 4, 1993, pp. 23-27.

p. 45: In other words ... their meaning. Blakey, Michael, "Socio-Political Bias and Ideological Production in Historical Archeology." In: Gero, Joan M., et al., *Socio-politics of Archeology*. (Department of Anthropology, Research Report No. 23.). Northhampton, Mass.: Amherst, 1983, p. 6.

p. 46: "I inserted myself ... he said." Cook, Karen, *op. cit.*

p. 47: Blakey went to ... Blakey exclaimed. *Ibid.*

p. 48: "There is no .... be reburied." General Services Administration, "Report to the [GSA] of the New York African Burial Ground Review Panel, May 12-14, 1993," p. 18.

p. 48: "[O]ur competitors, MFAT ... on ours." M. Blakey, "Dear Colleagues" Letter, July 10, 1992.

p. 49: "continuing actions," he ... New Yorkers." Cited in Cantwell, Anne-Marie, and Diana DiZerga Wall, *Unearthing Gotham*. New Haven: Yale, 2001. p. 286.

p. 49: "This is our ... as usual!" Cited by Barbara Stalling-Whited-Muniz. In: General Services Administration Region 2, "Comments on the Draft Research Design for Archeological, Historical, and Bioanthropological Investigations of the African Burial Ground...", New York: GSA Region 2, compiled comments. Comment 2, P 6.

p. 49: "Don't waste your ... your disrespect." Cantwell and Wall, *op. cit.*, p. 287.

GOING FORWARD — Chapter 4

p. 52: The excavated area ... of burials. ARCHEFR. p. 87.

p. 53: "To correct past .... Research Design." Dodson, Howard, Testimony to House Subcommittee on Public Buildings and Grounds, September 24, 1992.

p. 54: "But Dodson also .... Mayor's committee." Schomburg Center for Research in Black Culture, Minutes of Mayor's committee on the ABG, August 13. 1992. Courtesy of the Schomburg Center for Research in Black Culture.

p. 55: "For MFAT to ... Descendant Community." Cook, Karen, *Village Voice*, May 4, 1993, pp. 23-27.

p. 55: There is no ... the dead!'" *New York Times*, August 9, 1992, p. 45.

p. 55: "you hear all .... are people," *ibid.*

SANKOFA?

p. 55fn: The Descendant Community . . . an opinion. . . ." *Chronicle of Higher Education*, March 16, 1994, p. A10.

p. 56: "This is enough! . . . over them." *New York Times*, February 26, 1963, p. B13.

p. 56: "So many of . . . peace!" *New York Times, op. cit.*

p. 57: "It was almost . . . an inspection!" Cook, Karen, *Village Voice*, May 4, 1993, pp. 23-27.

p. 58: "He had this . . . that out," *ibid.*

p. 58: The Research Design . . . . human remains. U.S. Advisory Council on Historic Preservation, and University of Nevada, Reno, Introduction to Section 106 Review. Reno: University of Nevada, Reno, 2003. Tab 4.

p. 58fn: But Blakey acknowledged . . . good shape." *Archeology*, March/April 1993, p. 35.

p. 59: This draft explores . . . the authorities." Blakey, Michael, *Draft Research Design*, October 15, 1992, p. 13.

p. 60: "Unmolested by outside . . . . took shape," *ibid.*, pp. 14-15.

p. 60: "Blakey promises that . . . and evaluated," *ibid.*, p. 25.

p. 60: "The document promises . . . ethnic groups," *ibid.*, p. 53.

p. 62: "the remains are . . . . the project!" *Ibid.*, pp. 2-3.

TAKING OVER — Chapter 5

p. 63: When Blakey and . . . interested parties. Blakey, Michael, *Draft Research Design*, October 15, 1992.

p. 64: I see the . . . . naming ceremony! General Services Administration Region 2, "Comments on the Draft Research Design for Archeological, Historical, and Bioanthropological Investigations of the African Burial Ground..." New York: GSA Region 2, compiled comments. Letter, Carlton Reid to Peter A. Sneed.

p. 65: Many of these . . . . to science?" Letter, Abd'Alleh Letif Ali to Peter A. Sneed, *ibid.*

p. 66: "A less sophisticated . . . . the research!" Letter, Eloise W. Dicks to Peter A. Sneed, *ibid.*

p. 67fn: Project archeologists reported . . . Native Americans. ARCHEFR, p. 51fn.

p. 70: How could he . . . health changes! Blakey, Michael, *op. cit.*, p. 56.

p. 70: their "cranial morphology . . . . European descent." General Services Administration, Dec. 14, 1993, *op. cit.*, p. 20.

p. 70: "sufficiently complete for . . . . other . . . populations." Blakey, Michael, "The New York African Burial Ground Project; an examination of enslaved lives, a construction of ancestral ties," *Transforming Anthropology*, vol 7, 1998, p. 55.

p. 70: "Modern forensic science . . . are found." Letter, Alexander Sonek to Peter A. Sneed, General Services Administration Region 2, *op. cit.*

p. 71: "I am greatly . . . . physical anthropology?" Letter, Madeleine J. Hinkes to Peter Sneed, *ibid.*

p. 71: One of Hinkes' colleagues. . . are present." Letter, J. Michael Hoffman to Peter A. Sneed, *ibid.*

p. 72: A critical view . . . for adequacy." Letter, Jerry C. Rose to Peter A. Sneed, *ibid.*

p. 73: "Blakey rebutted such . . . her ancestry?" LaRoche, Cheryl and Michael L. Blakey, "Seizing Intellectual Power: The Dialogue at the New York African Burial Ground," *Historical Archeology*, vol 31 (No. 3), pp. 84-106, 1997.

p. 74: "I find the . . . in Canada." Letter, Jerry Melbye to Peter Sneed, General Services Administration Region 2, *op. cit.*

p. 74: Blakey would deny . . . human remains. ANTHROFR, p. 26.

COUNTERATTACK — Chapter 6

p. 75: Unfortunately, the chairman . . . present form. General Services Administration Region 2, "Comments on the Draft Research Design for Archeological, Historical, and Bioanthropological Investigations of the African Burial Ground..." New York: GSA Region 2, compiled comments. Letter, Laurie Beckelman to William J. Diamond, December 30, 1992.

p. 76: "Unfortunately," ACHP archeologists . . . . (ABG) site. Letter, Robert D. Bush to William Diamond, *ibid.*

p. 77: "On the basis . . . . this project. . . ." Letter, Nancy Demyttenaere to Howard Dodson, January 25, 1993. Courtesy of the Schomburg Center for Research in Black Culture.

p. 78: The purpose of . . . . professional anthropologist." University of Massachusetts-Amherst, "Anthropology Department Admission Guide for Admissions," 2005. pp. 6-8. This Department says its guidelines remained the same between 1979 and 2005. http://web.archive.org/web/20050226060316/http://www.umass.edu/anthro/graduate.html

p. 78: Blakey's thesis has . . . . the thesis. Blakey, Michael L., *Stress, Social Inequality, and Cultural Change: An Anthropological Approach to Human Psychophysiology*. PhD diss., University of Massachusetts, Amherst, 1985.

p. 79: "What can these . . . as saying,"*ibid.*, p. 103.
p. 79: "Their refusal skewed . . . . ethnic prejudices," *ibid.*, p. 104.
p. 80: "The purpose," he . . . economic control . . . *ibid.*, p. 137.
p. 82: Scientists at Lehman . . . president said. Letter, Ricardo R. Fernandez to David A. Paterson, February 5, 1993. Courtesy of the Schomburg Center.
p. 83: "We could take . . . of people!" Cook, Karen, *Village Voice*, May 4, 1993 pp. 23-27.
p. 83: In a letter . . . own hands. Letter, David A. Paterson to Ricardo R. Fernandez, January 18, 1993. Courtesy of the Schomburg Center.
p. 83: "As part of . . . American discovery." Letter, David Kurtz to Peter A. Sneed, General Services Administration Region 2, *op. cit.*
p. 84: Moore protested, saying . . . film's integrity. Letter, Christopher Moore to Howard Dodson, November 23, 1993. Courtesy of the Schomburg Center.
p. 85: "the study of . . . to change. *Downtown Express*, June 7, 1993.
p. 85fn: GSA paid for . . . related activities. Donald Eigendorf (GSA) to Author, May 20, 2009.
p. 86fn: But: Ninety percent . . . York City. *New York Times*, October 7, 2005.

SCIENCE AND INSPIRATION — Chapter 7

p. 87: GSA sought to . . . by letter, Letter, Robert W. Martin to Howard Dodson, May 17, 1993. Courtesy of the Schomburg Center for Research in Black Culture.
p. 88: Anthropologist Larsen already . . . fully adequate." General Services Administration Region 2, "Comments on the Draft Research Design for Archeological, Historical, and Bioanthropological Investigations of the African Burial Ground...". New York: GSA Region 2. compiled comments. Clark S. Larsen to Peter A. Sneed, November 23, 1992.
p. 90: "I am extremely . . . . New York." Letter, David Dinkins to Charles Rangel, May 21, 1993. Courtesy of the Schomburg Center for Research in Black Culture.
p. 92: The peer reviewers' . . . review of it. GSA, "Report to the GSA of the New York African Burial Ground Review Panel, May 12-14," 1993.

MOVING BODIES — Chapter 8

p. 98: Black people "should . . . . from that." Schomburg Center for Research in Black Culture, Transcript, A Public Forum on the Draft Proposal to the U.S. Congress for Commemorating the African Burial Ground. June 14, 1993, p. 173. Courtesy of the Schomburg Center for Research in Black Culture.

p. 99: In September, he . . . or attributions . . ." Lydia Ortiz to Michael Blakey, September 13, 1993 in ANTHROFR, p. 27.

p. 99: "In fact no . . . content existed," *ibid.*

p. 99fn: This racial selectivity . . . American professionals . . ." GSA, "Foley Square African Burial Ground Physical Anthropological Peer Review Panel Report," n.d. [1993], unpaginated.

p. 100: "I have no . . . . the remains." Esther Dawson, quoted in *Update* I(2), Fall, 1993, p 3.

p. 100: "It's a loss . . . about it." Claudia Milne, quoted in *ibid.*

p. 100: Another lab technician . . . our experience." Doville Nelson, quoted in ibid. p.4.

p. 101: Historian Christopher Moore . . . . Dr. Blakey. Christopher Moore, quoted in *ibid.* p. 4.

p. 101: "I fully respect . . . . to me." Miriam Francis, quoted in LaRoche, Cheryl and Michael L. Blakey, *Seizing Intellectual Power: The dialogue at the New York African Burial Ground,* Historical Archeology, vol 31 (No, 3), pp. 84-106, 1997.

p. 102: "Two stipulations to . . . be ascertained. *Foley Square African Burial Ground Physical Anthropological Peer Review Panel Report, op. cit.*

WHAT MIGHT BE FOUND — Chapter 9

p. 109: "[W]hat better way . . . . of life." Limp, W. Frederick and Jerry C. Rose. Introduction to Rose, J. C. (ed.), *Gone to a Better Land.* Arkansas Archeological, Survey Research Series, No. 25, 1985, p. 2.

p. 109: "Human bone," explains . . . . individual/population. Rankin-Hill, Lesley, A. *Biohistory of 19th-Century Afro-Amercians,* pp. 36-37

p. 111: "suggest that iron . . . . have been. Goodman, Alan H. and Debra A. Martin, "Reconstructing Health Profiles from Skeletal Remains," In: Steckel and Rose, *The Backbone of History.* Cambridge University Press, 2002, pp. 61-81.

DIGGING IN — Chapter 10

p. 116: The Research Design . . . successful conclusion." GSA: "Report

SANKOFA?

to the GSA of the New York African Burial Ground Peer Review Panel, May 12-14, 1993, p. 2.

p. 116: One panelist had . . . "first rate." *Chronicle of Higher Education*, March 16, 1994, p. A10.

p. 117: Howard began buying . . . November 1, 1996. Contract GS-02P-93-CUC-0071 and amendments, between GSA and Howard University.

p. 118: Howard University would . . . final illustrations." *Ibid.*, p. 47.

p. 121: "I essentially have . . . . is dominant." Mary C. Hill to Florence B. Bonner, May 28, 1996.

p. 122: A month later . . . diagnostic team." Mary C. Hill to Michael L. Blakey, June 5, 1992.

p. 123: "What you sent . . . to work!" Letter, Michael L. Blakey to Mary C. Hill, July 10, 1992.

A MUSKET BALL AND BEADS — Chapter 11

p. 128: One of the . . . . following year. Paleopathology Association Membership List, January, 1995; Paleopathology Association Supplemental Membership Directory, May, 1996.

p. 129: Later, Blakey explained. . .a Sankofa. Blakey, Michael, "Return to the African Burial Ground," www.archeology.org, accessed April 20, 2009.

p. 130: An art historian . . . at Howard . . . . *Update, Summer 1995* I(8), p. 3.

p. 132: "A new report said . . . the head." *Update, Fall 1994* I (5).

STUDYING THE BONES — Chapter 12

p. 134: This "generally poor . . . . life conditions." *Update, Spring 1995* I(7).

p. 136: But "there are . . . . the cemetery." ARCHEFR, p. 150.

p. 136: They "provide an . . . Mack says." *Update, Winter, 1995* I(6).

p. 136: "Our available . . . data . . . . our past. *Update, Spring, 1995, op. cit.*

p. 140: In this population . . . . excessive loads. Women, endurance, enslavement: exceeding the physiological limits. *Report to American Association of Physical Anthropologists*, 1995. See also, published abstracts for organized symposium, Martin, Debra L., organizer and chair.

p. 141: "We have no . . . Native American." Letter, M. Blakey to GSA, May 15, 1995.

p. 141: "Much later, it . . . . for Hill." Letter, Michael Blakey to Lydia

Ortiz, May 15. 1995.

LOGGERHEADS — Chapter 13

p. 142: "On December 22nd . . I've said." Hill, Mary C. Memo to self, December 22, 1994.

p. 144: On January 11, 1995 . . . . otherwise advised.' *Ibid.*, January 11, 1995.

p. 145: "My training in . . .final grade." Mary Hill to Florence B. Bonner, May 28, 1996.

p. 146: "forensics work relies . . . biocultural approach." ANTHROFR, p. 21.

p. 147: Blakey, a co-author . . . . She did. Unpublished draft, Women, endurance, enslavement; exceeding the physiological limits. American Association of Physical Anthropologists, 1995.

p. 148: trials following a . . . . been found. Lepore, Jill, *New York Burning*. Knopf, 2005, pp. 248-259.

p. 149: Today was my . . . . to tears. Mary Hill to Florence B. Bonner, June 1, 2005.

RACISM — Chapter 14

p. 153: Blakey, meanwhile, pridefully . . . . political concerns." LaRoche, Cheryl J., and Michael L. Blakey, Seizing Intellectual Power: The Dialogue at the New York African Burial Ground. *Historical Archeology*, vol. 31, (no. 3), pp. 84-106, 1997.

p. 154: "By omission northern . . . were denied." Blakey, Michael L., cited in *Current Biography Yearbook*, 2000, pp. 60-63.

p. 154: The "vindicationist" approach . . . . African-American culture . . ." LaRoche and Blakey, *op. cit.*, p. 89.

p. 154: "When vindicationist motivations . . . . and accuracy. LaRoche and Blakey, *op. cit.*, p. 91.

p. 155: "Thinking back over . . . . deliver them." Barbour, Warren T., Musings on a dream deferred. *Federal Archeology Report*, Spring, 1994, pp. 12-13.

p. 157: Meanwhile, Blakey's personal . . . . in 711 A.D. "Michael Blakey" in *Current Biography Yearbook*, 2000 p. 63.

p. 157fn: Blakey's insistence on . . . . blind alleys . . ." Cruse, Harold, *The Crisis of the Negro Intellectual*. New York: Morrow, 1967, republished by New York Review of Books, 2005, p. 255.

WAR — Chapter 15

SANKOFA?

p. 160: Bonner mediated the ... to her ." Memo, Florence Bonner to Clarence M. Lee, June 13, 1996.
p. 162: By then Blakey ... Cobb lab. Contract GS-02P-93-CUC-0071, modification PC 14, January 31, 1995.
p. 164: This problem was .... among researchers." ANTHROFR p. 143.
p. 164fn: "is not qualified .... the data." Letter Mary C. Hill to Frank M. Caprio and Russell L. Sandidge, March 12, 1998.
p. 166: a " leading model ... 21st century." LaRoche, Cheryl and Michael L. Blakey, Seizing Intellectual Power: The dialogue at the New York African Burial Ground, *Historical Archeology*, vol 31 (No, 3), pp. 84-106, 1997.
p. 166: He spoke there ... blacks' struggle. Blakey, Michael L., The New York African Burial Ground Project: Examination of Enslaved Lives, a construction of ancestral ties. *Transforming Anthropology*, vol. 7 (no. 1), 1998, p. 55. Revised text of UN speech.

DATA AND DISCORD — Chapter 16

p. 168: He also sent ... and others." *Update* II(8), Fall 1998, p. 11.
p. 169: "Given [that] the ... individually belong." ANTHROPR, p. 35.
p. 169: "Given the high ... record alone," *ibid.*, p. 53.
p. 170: "We suspect, moreover, ... more prevalently," *ibid.*, p. 100.
p. 172fn: "Your statement" that ... egregious misrepresentation." Letter, Robert W. Martin to Michael Blakey, December 23, 1998.
p. 174fn: "[G]enetic data from ... genetic relationship." [Blakey, Michael L], Howard University and John Milner Associates, *Drafted Research Design for Archeological, Historical, and Bioanthropological Investigations of the African Burial Ground*, prepared for General Services Administration, Region 2, April 22, 1993. pp. 68-69.
p. 176: "all reports of .... actual reports." Letter, Michael Blakey to Lisa Wager, March 5, 1999.
p. 177: "which stipulated that ..... level of effort." General Services Administration, Region 2. *Research Design for Archeological, Historical, and Bioanthropological Investigations of the African Burial Ground*, New York, N.Y., December 14, 1993.

REJECTION — Chapter 17

p. 178: On August 12, ... rejected it. Letter, William Lawson to Michael Blakey, August 12, 1999, as quoted by Michael Blakey to William Lawson (see immediately below).

p. 179: "The reviewers are .... completed draft. Letter, Michael Blakey to William Lawson, December 31, 1999.

GENETIC FOREBEARS — Chapter 18

p. 187: In a scientific.... African origins." Kittles, Rick, et al., Genetic Variation and Affinities in the New York Burial Ground of Enslaved Africans. *American Journal of Physical Anthropology*, Supplement 28, 1999. p. 170.

p. 187: "I knew black .... to Africa. Ford, Sam: "His DNA Promise Doesn't Deliver," *Los Angeles Times*, May 29, 2000.

p. 188: "This wasn't clear-cut .... come from," *ibid*.

p. 188: "I know that .... just curious!" "Flesh and blood and DNA," http://www.salon.com/2000/05/12/roots_2/.

p. 188: "I did the .... to Germany!" Cited in *Los Angeles Times*, op. cit.

p. 189: "help restore the ... less human." *New York Times*, August 28, 2000.

p. 189: "It really ... to end!" *ibid*.

p. 191: "Unfortunately," GSA official .... [the] schedule." *Update* III(2), Winter 2000, p. 13.

p. 192: The chief of .... Parsons said." *New York Times, op. cit.*

p. 192: "To a lot .... before that." *Boston Globe*, April 13, 2000.

p. 194: Jackson objected, as .... God-given right!" Salon.com, op. cit.

GSA TAKES OVER — Chapter 20

p. 207: They set as .... July 31, 2002. *Activity Report African Burial Ground Project*. July 31, 2002, p. 5.

p. 208: Before the March ... professional editor, *ibid.*, p. 4. All subsequent quotes in Chapter 20 are from this Activity Report.

BETRAYAL — Chapter 21

p. 213: A month later .... the work. *Activity Report African Burial Project*, October 31, 2002, p. 4.

'COMMUNITY' VIEWS — Chapter 22

p. 217: The gathering ... examine" it. Miles, Steve, Convening Wel-

SANKOFA?

come. Transcript of the conference, African Geneology & Genetics, June 21, 2002. Minneapolis: University of Minnesota Center for Bioethics. http://web.archive.org/web/20060722021118/http://www.bioethics.umn.edu/afrgen/html/conveningwelcome.html

p. 218: "There is a .... no role." Graves, Joseph, Jr., "The Myth of Race: America's Original Science Fiction," *ibid.*, p. 130.

p. 218: "Science is a .... community level." powell, john, Opening Remarks, *Ibid.*, p. 132.

p. 218: The conference's co-sponsor .... and living." Azzahir, Atum, Convening Welcome, *Ibid.*, pp. 223-4.

p. 219: Slavery, attorney powell .... particular populations." powell, john, *op. cit.*, p.211.

p. 219: "I am ambivalent ... construct geneologies": Duster, Troy, "Tracing Lineage: A Social Project and a Genetic Stamp of Approval," *ibid.*, p.6.

p. 220: "the danger of ...., ethnic origin ... *ibid.*

p. 220fn: This situation appears .... white Europeans. Lamason, Rebecca L., et al. SLC24A5, a putative cation exchanger affects pigmentation in zebrafish and humans." *Science* 310, pp. 1782-6, December 16, 2005.

p. 221: "the challenge is .... own identity." Kittles, Rick, Transcript, *op. cit.*, pp. 16-22.

p. 223: "*Biological* races are .... heart attack." Graves, Joseph, *op. cit.*

p. 223: "Race is a ... social fact." powell, john, *op. cit.*, p. 106.

p. 223: "These [genetic] tests .... exact certainty." Bolnick, Deborah A., and Duana Fullwiley, Troy Duster, et al., "The Science and Business of Genetic Ancestry Testing." *Science* 318: 399-400, October 19, 2007.

FINISH LINE! — Chapter 23

p. 227: There are 248 ... follow-up studies. Contract between GSA and Howard University, Contract GS-02P-93-CUC-0071, Modification PC31, pp. 9-10, October 24, 2005.

p. 228: Besides the reburied ... of fill, ARCHEFR, p. 104.

p. 228: Two-and-a .... in 1993). http://web.archive.org/web/20120721214914/http://www.africanburialground.gov/ABG_FinalReports.htm.

p. 229: "[T]he Sankofa symbols .... that symbol." Blakey, Michael, "Return to the African Burial Ground," www.archeology.org, accessed April 20, 2009.

p. 229: The project's GSA . . . . class memorial . . . ." Greenberg, Alan, *Confessions of a Government Man*. Indianapolis: Dog Ear Publishing, 2010, p. 58.

p. 231: "We think the . . . . the fold." ARCHEFR, p. 155.

WHAT THEY FOUND — Chapter 24

p. 233: Only 48 sets . . . . not done. George, Matthew, *Progress Report: The African Burial Ground Project, Mitochondrial DNA Genetics Sub-Project*, n.d. [1995], offset.

p. 235: "[T]he quantity and . . . including Blakey. ANTHROFR, p. 268.

p. 236: One of these . . . . demographic analysis," *ibid.*, p. 270.

p. 236: "almost impossible to . . . aggregate data," *ibid.*, p. 271.

p. 236fn: This is the . . . . age two. ANTHOFR., p. 272.

p. 237: "metabolic disturbances of . . . the body." *Ibid.*, p. 306.

p. 237: "Hypoplasia [thus] provides . . . . general stress.'" *Ibid.*, p. 309.

p. 238: "the information presented . . . . these findings." *Ibid.*, p. 400.

p. 238: "strenuous labor started . . . . they work," *ibid.*, p. 448.

p. 239: "does not provide . . . . social milieu," *ibid.*, p. 512.

p. 240: "racist and inhumane . . . . shadowy evidence," *ibid.*, p. 542.

p. 241: "that the comparison . . . . seem overstated . . . ." Rothstein, Edward, "The Peculiar Institution as Lived in New York," *New York Times*, October 7, 2005.

IS IT SCIENCE? — Chapter 25

p. 244: "The ABG data . . . was ignored", GSA, *Foley Square African Burial Ground Physical Anthropological Peer Review Panel Reports*, n.d. [1993].

p. 244: The Final Report . . . Howard University. ANTHROFR, p. 297.

p. 246: "These findings indicate . . . forced migration." Howard University, *op. cit.*, p. 331.

p. 246: "These results provide . . . . in bondage," *ibid.*, p. 350.

p. 246: "The information presented . . . the ABG," *ibid.*, p. 400.

FINDINGS — Chapter 27

p. 249: to reconstruct the . . . of colonial-era Africans," *Update* I(2), Fall, 1993, p. 1.

p. 253: "We seek to . . . Western World." Cited in Slocum, Marcia,

SANKOFA?

Washington Post, August 27, 2002, p. 1.

p. 253: "I had the . . . . to proceed . . . ." *Newsday*, Queens, Long Island edition, April 4, 2000, p. 8.

p. 253: At a religious . . . . us forever." Cited in Heilen, Michael, *The Skeletal Biology, Archaeology, and History of the New York African Burial Ground: A Synthesis of Volumes 1,2, and 3* (Washington, D.C.: Howard University, 2004), p. 22.

p. 253: "[T]he skeletons that . . . to me." *Update* II(1), Winter, 1996-7, p. 15.

p. 254: "You have to . . . the research." *Update, op. cit.*, p. 4.

EPILOGUE

p. 257: the final, final . . . . late developments. Heilen, Michael, *The Skeletal Biology, Archaeology, and History of the New York African Burial Ground: A Synthesis of Volumes 1,2, and 3* (Washington, D.C.: Howard University, 2009), p. 3.

p. 259: There is no . . . . African design. Seeman, Erik R., *Death in the New World*. Philadelphia: U. Penn, 2010, pp. 207-215.

p. 259: "[T]he deceased were . . . New Yorkers." HISTFR, p. 184.

p. 261: "The research . . . was . . . . ever conducted." Heilen, Michael, *op. cit.*, p. 3.

p. 261: the project's long-time . . . . the checks." Greenberg, Alan L., *Confessions of a Government Man*. Indianapolis: Dog Ear Publishing, 2010, pp. 58-60.

DAVID ZIMMERMAN

BIBLIOGRAPHY

African Burial Ground Project [Michael L. Blakey, et al.], "Skeletal Biology Report," First Draft. New York: General Services Administration, Region 2, August 5, 1998, 3 vol.

Berlin, Ira, and Leslie M. Harris, eds., *Slavery in New York*, New York: New Press, 2005. Published in conjunction with the New York Historical Society.

Birmingham, Robert A. and Leslie E. Eisenberg, *Indian Mounds of Wisconsin*. Madison: University of Wisconsin Press, 2000.

Blakey, Michael L., "Stress, Social Inequality, and Cultural Change: An Anthropological Approach to Human Psychophysiology." PhD diss., University of Massachusetts, Amherst, 1985.

———. "Research Design for Temporary Curation and Anthropological Analysis of the 'Negro Burying Ground' (Foley Square) Archeological Population at Howard University." June 11, 1982.

———. *Draft Research Design*, October 15, 1992.

———. *Research Design*, April 22, 1993.

Blakey, Michael L., and Lesley M. Rankin-Hill, eds., *Skeletal Biology Final Report*. http://web.archive.org/web/20120721214914/http://www.africanburialground.gov/ABG_FinalReports.htm

Cantwell, Anne-Marie, and Diana Wall diZerega, *Unearthing Gotham*. New Haven: Yale, 2001. DiZerega

Condon, Cynthia, et al., *Freedman's Cemetery*. Austin, TX.: Texas Department of Transportation, Environmental Affairs Division, Archeology Studies Program, Report Number 9, 1998.

Cruse, Harold, *The Crisis of the Negro Intellectual*. New York: Morrow, 1967. Reprinted with preface by Stanley Crouch, New York: New York Review of Books, 2005.

General Services Administration, "Foley Square African Burial Ground Physical Anthropological Peer Review Panel Report," n.d. [1993].

General Services Administration, Region 2. Draft "Research Design for Archeological, Historical, and Bioanthropological Investigations of the African Burial Ground", New York, N.Y., Oct. 15,

1992. (Prepared by Howard University and John Milner Associates).

General Services Administration, Region 2. "Research Design for Archeological, Historical, and Bioanthropological Investigations of the African Burial Ground," April 22, 1993. (Prepared by Howard University and John Milner Associates.)

General Services Administration, Region 2. "Research Design for Archeological, Historical, and Bioanthropological Investigations of the African Burial Ground," New York, N.Y., Dec. 14, 1993, 131 pages plus unpaginated resumes.

_____.Late changes with new material, no date.

Goode-Null, Susan Kay, "Slavery's Children: A Study of Growth and Childhood. Sex Ratios in the New York African Burial Ground." PhD diss., University of Massachusetts, Amherst, 2002.

Gore, Al, *The Assault on the Reason*. New York: Penguin, 2007.

Hall, Gwendolyn, *Slavery and African Ethnicities in the Americas*. Chapel Hill: University of North Carolina Press, 2005.

Hansen, Joyce, and Gary McGowan, *Breaking Ground, Breaking Silence*. New York: Henry Holt, 1998.

Heilen, Michael, ed., *The Skeletal Biology, Archeology, and History of the New York African Burial Ground: A Synthesis of Volumes 1, 2, and 3* (Washington, D.C.: Howard University, 2009).

_____. *New York African Burial Ground: Unearthing the African Presence in Colonial New York* (Washington, D.C.: Howard University, 2009).

Hill, Mary Cassandra, "Porotic Hyperostosis as an Indicator of Anemia: An Overview of Correlation and Cause," PhD diss., University of Massachusetts, Amherst, 2001.

Howard University and John Milner Associates, "Draft Research Design for Archeological, Historical, and Bioanthropological Investigations of the African Burial Ground," prepared for General Services Administration, Region 2, New York, October 15, 1992.

_____."Revised Research Design for Archeological, Historical, and Bioanthropological Investigations of the African Burial Ground (Broadway Block) New York, New York," prepared for General Services Administration, Region 2, New York, April 22, 1993.

Kittles, Rick, *et al.*, "Genetic Variation and Affinities in the New York Burial Ground of Enslaved Africans," *American Journal of Physical Anthropology*, Supplement 28, 1999, p. 170.

LaRoche, Cheryl, and Michael L. Blakey, "Seizing Intellectual Power: The Dialogue at the New York African Burial Ground," *Historical*

Archeology 31(3), 1997, pp. 84-106.

Lepore, Jill, *New York Burning*. New York: Knopf, 2005.

Marable, Manning, *Malcolm X*. New York: Viking, 2011.

Medford, Edna Greene, *History Final Report*. New York: General Services Administration, Region 2, Nov., 2004. http://www.africanburialground.gov/ABG_FinalReports.htm

Olson, Steve, *Mapping Human History*. Boston: Houghton, Mifflin, 2003 (paperback).

Perry, Warren, Jean Howson and Barbara Bianco, eds., *Archeology Final Report*. New York: General Services Administration, Region 2, February 2006. http://www.africanburialground.gov/ABG_FinalReports.htm

Rankin-Hill, Lesley M., *A Biohistory of 19th Century Afro-Americans*. Westport, CT: Bergin & Garvey, 1997.

Roch, Mary, *Stiff*. New York: W.W. Norton, 2004 (paperback).

Schama, Simon, *Rough Crossings, Britain, The Slaves and the American Revolution*. New York: HarperCollins, 2006.

Seeman, Erik R., *Death in the New World: Cross-Cultural Encounters, 1492-1800*. Philadelphia, PA: U. Penn, 2010.

Shorto, Russell, *The Island at the Center of the World*. New York: Vintage, 2005.

Steckel, Richard, and Jerry Rose, eds., *The Backbone of History*. Cambridge: Cambridge University Press, 2002.

U.S. Advisory Council on Historic Preservation, and University of Nevada, Reno, *Introduction to Section 106 Review*. Reno: University of Nevada, Reno, 2003.

Williams, Juan, *Enough*. New York: Crown, 2006.

SANKOFA?

PEOPLE IN THIS BOOK

ABG = African Burial Ground, an 18th century black cemetery in Manhattan
ACE = Army Corps of Engineers
GSA = General Services Administration, a US federal agency
HCI = Historic Conservational and Intrepretation, Inc.
MFAT = Metropolitan Forensic Anthropology Team
Ali, Ab'dAlleh Letif — New York sheik who questioned the scientific benefits of an anthropological study of ABG's human remains.
Armelagos, George J., PhD — Michael Blakey's advisor at UMass Amherst.
Austin, Richard G., PhD — GSA chief in Washington, DC.
Azzahir, Atum — Leader of the Powderhorn/Phillips Cultural Wellness Center.
Bannon, Charles — Member of Descendant Community.
Barbour, Warren T.D., PhD — Associate director of the ABG project.
Bass, William, PhD — Forensic anthropologist at the University of Tennessee who trained Cassandra Hill.
Beckelman, Laurie — Chairman of NYC Landmarks Preservation Commission
Blakey, Michael L., PhD — Anthropologist and scientific director of research on the ABG.
Blakey, Tariq — Michael Blakey's son.
Bonner, Florence B., PhD — Chair of Howard University's Sociology-Anthropology Department.
Brock, Donna - Howard University spokeswoman.
Broughton, Mildred - Contracts and Freedom of Information officer at GSA.
Brown, Emilyn - Editor at Update newsletter.
Bush, Robert D., PhD — Executive Director of the Federal Advisory Council on Historic Preservation.
Butler, Carrie — Maid in home of segregationist Strom Thurmond, whom he impregnated.
Cantwell, Anne-Marie, PhD — NYC archeologist and writer.
Chapman, William, T., MD — Would-be tourist to Africa.

Cobb, W. Montague, MD — Physician and physical anthropologist who fought racism in the mid 20th century. He started a collection of skeletal remains, mostly of black people, at Howard University in Washington, D.C.

Cole, Johnnetta B., PhD -- Member of Michael Blakey's thesis committee at UMass Amherst and his supporter.

Cole, O. Jackson, Jr., PhD — Howard University executive in charge of ABG research.

Cook, Karen — Village Voice reporter.

Collier, Melvin — Would-be tourist to Africa.

Condon, Keith, PhD - Biologist, Indiana University School of Medicine.

Cowen-Ricks, Carrel, PhD — Archeologist, Clemson University; reviewer of Michael Blakey's Research Design.

Crick, Francis. PhD– Researcher who identified (with James Watson) the double-helical structure of DNA.

Cruse, Harold — NYC black intellectual and social critic, mid-20th century.

Davis, Thurman M., Sr. — GSA administrator.

Dawson, Esther — Interviewee who favored sending cemetery remains to Howard University for study.

DeCorse, Christopher R., PhD– Archeologist and expert on African-made beads.

Demyttenaere, Nancy, MA — Archeologist, a GSA consultant, who was highly critical of Michael Blakey's mien and manner.

Diamond, William — GSA regional administrator in the early 1990s.

Dicks, Eloise W. — Vocal member of the Descendant Community.

Dinkins, David N. — NYC Mayor, 1989-93; city's first and only black mayor.

Dodson, Howard, MA — Historian and New York Public Library official, who led Black Community campaign for respectful treatment of ABG's remains.

Donaldson, James A., PhD — Mathematician and Howard University official.

Dunston, Georgia G., PhD — Molecular biologist and co-director of the National Human Genome Center at Howard University.

Duster, Troy, PhD — Sociologist at New York University, NYC.

Eisenberg, Leslie E., PhD — Anthropologist who helped excavate ABG.

Engerman, Stanley — Historian of slavery.

Fogel, Robert — Historian of slavery.

Ford, Sam — Washington, D.C., TV reporter.

Foreman, Sylvia H., PhD - Chairman of the UMass Anthropology

SANKOFA?

Department and of Michael Blakey's thesis committee.
Francis, Miriam – Artist, activist, and Steering Committee member.
Fraunces, Samuel — Eighteenth-century black tavern-keeper in NYC, who was an aide to President George Washington.
George, Matthew, PhD — Howard University biochemist and molecular biologist.
Goldberg, Carey — New York Times reporter.
Goode-Null, Susan K., PhD – Osteologist, who as graduate student analyzed ABG bones
Goodman, Alan H., PhD — Anthropologist at Hampshire College.
Graves, Joseph A. Jr., PhD — Evolutionary biologist, Arizona State University, Pheonix.
Greene, Marcia Slocum — Washington Post reporter.
Haley, Alex — Black author in the 20th century who wrote — not always accurately — about slavery.
Hall, Gwendolyn Midlo, PhD — Black historian who unearthed slave genealogies in Louisiana courthouse attics.
Hansen, Joyce — Black author of children's books.
Hatim, Muhammad – Representative of the Admiral Family who worried about ABG remains.
Heberley, Lori, MA – Howard therapist who met with Mary Cassandra Hill.
Hemings, Sally — Thomas Jefferson's slave and lover.
Henderson, Cassandra — GSA public relations consultant.
Hill, Mary Cassandra — Osteologist for the ABG project at Howard University.
Hinkes, Madeleine, J., PhD — Harsh critic of Michael Blakey's research plan.
Hoffman, J. Michael, MD — Strong critic of Michael Blakey's research design.
Howson, Jean, PhD — Co-principal archeologist for the ABG project.
Ingrassia, Robert — Reporter for New York Daily News who exposed ABG problems.
Jackson, Brian A. — GSA official.
Jackson, Bruce, PhD — Boston molecular biologist and creator of forensic DNA program.
Jackson, Fatimah L.C., PhD — ABG molecular biologist.
Jackson, John A. — Howard University attorney.
Karklins, Karlis — Archeologist, expert on African beads.
Keita, S. O. Y., MD — Howard University surgeon with an interest in anthropology.
King, Rodney — Black Los Angeles motorist badly beaten by police in 1991.

King, Virginia H. — Historian of slavery.
Kiple, Kenneth F. — Historian of slavery.
Kittles, Rick, PhD — Geneticist at Howard University.
Kurtz, David L. — Filmmaker for the ABG project.
LaRoche, Cheryl J., PhD — Archeologist at John Milner Associates who studied beads exhumed at ABG; sometime co-author of reports by Michael Blakey.
Larsen, Clark S., PhD — Anthropologist at Purdue University; reviewer of Michael Blakey's Research Design.
Law, Ronald — GSA's associate regional administrator overseeing the ABG project.
Lawson, William — Official at GSA regional headquarters in New York.
Levitt, Norman, PhD — NYC science critic and historian, deceased.
Mack, Mark E. — Lab director at Howard University for the study of ABG remains.
Maddox, Alton H., Jr. — Lawyer and Black Community leader in NYC.
Martin, Debra L., PhD — Anthropologist at Hampshire College.
Martin, Robert W. — GSA administrator in New York City.
McGowan, Gary S. — Conservator for ABG project artifacts.
McManamon, Frank, PhD — National Park Service archeologist who volunteered to be a peer reviewer for Blakey's Skeletal Biology Final Report.
Medford, Edna G., PhD — An Author of ABG History Final Report.
Miles, Steven, MD — Physician and bioethicist, University of Minnesota, in Minneapolis.
Milne, Claudia — Graduate student who supported sending human remains to Washington for study.
Miscione, Renée — GSA's circumspect public relations official.
Moore, Christopher — Historian and writer of ABG film.
Nelson, Doville — Lab technician who opposed sending human remains to Howard University for study.
Ofori-Ansa, Kwaku, PhD — Art historian at Howard University.
Ortiz, Lydia — GSA official in Washington, D.C.
Parrington, Michael, PhD — Principal archeologist for HCI; deceased.
Parrish, Silas L., MSW — Mental health consultant for the District of Columbia.
Parsons, Thomas J., PhD — Geneticist and chief of research at the Armed Forces DNA identification laboratory.
Paterson, David A. — Black community leader and New York State senator, and later governor of New York.
Paynter, Robert, PhD — Professor of anthropology at UMass Amherst

who was a student there with Blakey in the 1980s.
Perry, Stephen A. — GSA administrator.
Perry, Warren, PhD — Co-principal archeologist for the ABG project.
Pieterszen, Abraham — Dutch settler who owned land adjacent to ABG.
Pinkett, Mary — Member of Descendant Community
Pointer, Noel — Jazz musician (deceased) who fought to save ABG and its human remains.
powell, john a., JD - Chief, Institute of Race and Poverty, University of Minnesota.
Ramsey, Eleanor Mason, PhD — Anthropologist; reviewer of Michael Blakey's Research Design.
Rabinowitz, Richard, PhD - Curator for New York Historical Society.
Rangel, Charles — Congressman (Dem., New York) and strong supporter of the ABG project.
Rankin-Hill, Lesley, PhD — UMass anthropology alumna who was co-editor of skeletal biology preliminary and final reports; a close colleague of Michael Blakey.
Rathbun, Theodore (Ted) A., PhD — Anthropologist who disinterred a black cemetery in South Carolina for study; reviewer of Michael Blakey's Research Design.
Reid, Carlton — Fought for ABG's preservation.
Rodriguez, Jan — Black fur trader who was apparently the first settler in Manhattan.
Rose, Jerome (Jerry) C., PhD — Anthropologist, GSA's monitor of research on the ABG remains; reviewer of Michael Blakey's Research Design.
Rothstein, Edward — *New York Times* writer.
Rutsch, Edward S. — Archeologist from HCI who investigated the ABG project; deceased.
Sandidge, Russell L. — Alabama attorney for Cassandra Hill.
Savage, Augustus (Gus) — Black Chicago congressman (Dem., Illinois), who stopped the destruction of the ABG site.
Schomburg, Arthur A. — New York black scholar who articulated the vindicationist viewpoint.
Sciulli, Paul W., PhD — Developer, with Jerry Rose and Richard Steckel, of Health Index database derived from skeletal remains.
Scorcia, John — GSA official.
Sneed, Peter — Urban planner at GSA's New York regional headquarters who worked on ABG project.
Sonek, Alexander, PhD — Forensic anthropologist critical of Michael Blakey's Research Design.

Stampp, Kenneth — Historian of slavery.
Staples, Brent — New York Times writer.
Steckel, Richard H., PhD — Developer, with Jerry Rose and Paul W. Sciulli, of Health Index database derived from skeletal remains.
Sullivan, Margaret V. — Human Resources Official at Howard University.
Sutphin, Amanda, RPA, — New York City Archeologist.
Swedlund, Alan C., PhD — Member of Michael Blakey's thesis committee at Amherst UMass
Swygert, H. Patrick — President of Howard University.
Tabasi, Adunni Oshupa — NYC black activist.
Taylor, James V., PhD — Early anthropologic investigator of the ABG; MFAT co-director.
Taylor, Rodger — Interviewer and writer.
Terry, Esther, PhD— Afro-American literature professor and associate chancellor of UMass Amherst who knew Blakey when he studied there starting in the late 1970s.
Thomas, R. Brooke, PhD — Co-chairman of Michael Blakey's thesis committee at UMass Amherst.
Thurmond, Strom — Segregationist US senator who impregnated black maid Carrie Butler when he was a young man and she was a teenager.
Tolbert, Emory J., PhD — Howard University historian.
Trimble, Michael K. (Sonny), Jr., PhD — ACE forensic anthropologist who led ABG project from 2002.
Turkel, Spencer, PhD — NYC anthropologist in charge of ABG excavation and study in 1991 and 1992; MFAT co-director.
Wade, George W., PhD — Member of Michael Blakey's thesis committee at UMass Amherst.
Wager, Lisa — GSA's executive director of ABG project.
Walken, Christopher — Movie actor who portrayed the headless horseman in the film based on Washington Irving's short story.
Walker, Philip L., PhD — Anthropologist at the University of California at Santa Barbara; reviewer of Michael Blakey's Research Design.
Wall, Diane diZerega, PhD — NYC archeologist and author.
Watson, James. PhD - Researcher who identified (with Francis Crick) the double-helical structure of DNA.
Weis, Theodore (Ted) — Liberal NYC congressman in the late 20th century, for whom the federal office building on the ABG site was named.
Williams, Gary — Member of Descendant Community.
Wilson, Sherrill D., PhD — Head of ABG project's Office of Public

SANKOFA?

Education and Interpretation; editor of Update.

Woodson, Carter G. — Black activist-scholar who formulated doctrine of vidicationism.

Wright, Howard — NYC black activist.

Yancy, Rodney — Official at Thomson Scientific Company in Philadelphia.

Yarbrough, Cathy — Public relations official for National Human Genome Research Institute.

DAVID ZIMMERMAN

TIME LINE

This outline is based in part on a "Project Chronology and Timeline" prepared by the African Burial Ground project Office of Public Education and Interpretation (OPEI). (http://web.archive.org/web/20060206172938/http://www.africanburialground.gov/OPEI_Documents/OPEI_Project_Chronology.htm.

ABG = African Burial Ground, an 18th century black cemetery in Manhattan
ACE = Army Corps of Engineers
ACHP = Advisory Council on Historic Preservation
GSA = General Services Administration, a US federal agency
HCI = Historic Conservational and Intrepretation, Inc.
MFAT = Metropolitan Forensic Anthropology Team
MOA = Memorandum of Agreement (MOA)

1626
First black slaves arrive in New Amsterdam (New York).

1697
Trinity Church closes its graveyard to blacks (and Catholics and Jews), who must now bury their dead elsewhere.

1700
An estimated 700 blacks live in Manhattan, 14 percent of the population.

1712
Estimated time of first burials in "Negroes Burial Ground" outside town palisade. Cemetery is later renamed.

1795
Cemetery closed. In 19th and 20th-centuries site is overbuilt with homes and businesses; cemetery is largely forgotten.

SANKOFA?

1989
GSA prepares to build federal office building on site and signs Memorandum of Agreement (MOA) with Advisory Council on Historic Preservation (ACHP) that provides guidelines for protecting ABG.

January 1991
Cultural and archeological surveys of site begin, as required by law.

May 1991
Human remains discovered.

September 1991
Excavation of human remains begins. Mayor David Dinkins conveys Black Community's concerns about site's degradation to GSA.

October 8, 1991
At press conference on site, black leaders express outrage at GSA's cavalier treatment of ABG human remains.

December 9, 1991
Black legislator establishes task force for oversight of ABG. It includes wide cross section of concerned citizens, particularly members of Black Community.

December 20, 1991
New MOA signed between GSA, ACHP, and New York City Landmarks Preservation Commission to further protect site and mitigate damage and destruction.

January 7, 1992
At GSA public forum, anthropologist Michael L. Blakey presents "Research Design for Archeological, Historical, and Bioanthropological Investigations of the ABG." Human remains would be removed from New York to Washington, D.C.

February 14, 1992
Construction workers damage several graves. Inaccurate maps are blamed.

David Zimmerman

March 1992
Small New Jersey salvage archeology firm, Historic Conservation and Interpretation, Inc. (HCI), submits research design for site to GSA. HCI works in tandem with Metropolitan Forensic Anthropology Team (MFAT) researchers who are disinterring and then storing cemetery's black human remains.

March 17, 1992
GSA authorizes documentary film on ABG for school audiences.

April 23, 1992
Mayor Dinkins sets up Advisory Committee on the ABG project.

July 1992
ACHP nixes HCI draft research design. Blakey and Howard University, an historic black school in Washington, D.C., submit alternative research design.

July 1, 1992
John Milner Associates (JMA), an archeological firm in West Chester, Pennsylvania, assumes administrative control of ABG work.

July 27, 1992
At congressional hearing near ABG, Mayor Dinkins charges that GSA violated the MOA. Congressman Gus Savage, whose subcommittee has oversight, threatens GSA, which caves in and agrees to temporarily halt further excavation.

July 29, 1992
Work stopped by GSA. About 400 burials have been removed; a dozen remain exposed.

August 1992
Blakey condemns MFAT's poor care of skeletal remains.

September 1992
GSA opens archeological lab in World Trade Center under contract with JMA.

October 9, 1992
Excavation of exposed burials completed; archeological site closed so office building construction can proceed.

SANKOFA?

October 15, 1992
Blakey and JMA submit new draft research plan.

October 26, 1992
Congress creates Federal Steering Committee (FSC) for ABG.

November 1992
MFAT members urge colleagues to write GSA and oppose Blakey research plan.

November 1992
Startup money, $3 million, allocated by Congress in bill introduced by New York Senator Alphonse D'Amato (Rep.), passed, and signed by President George H. W. Bush.

December 1992
FSC says cemetery remains should go to Howard University, thereby favoring Blakey and cutting off MFAT.

January 1993
Professional colleagues critique Blakey's research plan in letters to GSA. Few favor his controlling the research.

February 25, 1993
New York Landmarks Preservation Committee designates ABG area a New York Landmark, opening way for its designation as a National Historic Landmark by Secretary of Interior and Congress in April.

April 22, 1993
Revised Research Design submitted to GSA by Blakey, Howard, and JMA.

May 20, 1993
ABG Office of Public Education and Interpretation opens in World Trade Center to provide public outreach for ABG project.

August 6, 1993
FSC sends final recommendations to Congress on ABG. Blakey wins all. MFAT loses out.

August 12, 1993
Contract signed by GSA and Howard starts money flow for research. First payment, to move bones to Washington: $261,481.

David Zimmerman

September 13, 1993
First of human remains sent from MFAT to Blakey at Howard. Others soon follow.

August 1994
"Firm fixed price" for work at Howard now $2,828,000. Blakey on payroll at $250 per diem. Research staff now includes Lesley Rankin-Hill, associate director; Mark Mack, laboratory director; and Mary Cassandra Hill, osteologist (bone expert). Skeletal remains are cleaned, laid out, mended, measured, and examined.

September 30, 1994
GSA disbands FSC.

December 1994
Blakey and Hill at loggerheads over scientific, personnel, and personal issues.

Winter 1994 to Summer 1995
Blakey-Hill conflict worsens.

January 31, 1995
Fourteenth modification of GSA-Howard contract brings total to $3,612,264, or close to $10,000 per set of remains.

Summer 1995
Update newsletter of ABG reports Blakey awarded honorary degree by branch of City University of New York. It cites him as "a good role model" for students considering graduate studies in anthropology.

March 1995
Blakey, Mack, and Hill present first — and virtually only — oral scientific reports on ABG science to colleagues at American Association of Physical Anthropologists conference in Oakland, California.

March 1995
Blakey and Hill at loggerheads over scientific, personnel, and personal issues.

August 31, 1995
Blakey fires Hill.

SANKOFA?

April 1, 1996
Contractual deadline for first draft of preliminary bio-anthropological report passes.

August 1996
Blakey presents ABG work to United Nations subcommittee on human rights in Geneva, Switzerland. He says ABG research constitutes new dimension in "long-standing human rights struggle among African-Americans."

August 5, 1998
Blakey submits first draft of *Skeletal Biology Report* to GSA. Costs now exceed $13,000 per set of human remains.

August 12, 1999
After a year, GSA rejects first draft as unsatisfactory.

December, 1999
Blakey responds by letter to GSA rejection. GSA orders audit and tells Howard University to lock the doors of Blakey's lab. Blakey goes to Brown University in Providence, Rhode Island, for a year.

February 2001
Newspapers report ABG debacle for first time.

September 11, 2001
ABG archeological lab in World Trade Center destroyed in Al-Queda attack. Most specimens later recovered. Meanwhile, relations between Blakey and GSA reach their nadir.

March 2002
GSA seizes control of ABG project, appoints US Army forensic anthropologist Michael Trimble, PhD, to take over. Other federal agencies, including National Park Service, agree to help bring project to closure. Blakey had complained that Cassandra Hill's results were unacceptable. Trimble and colleagues say her work is fine and move forward to use it to complete the *Final Report*.

August 27, 2002
Washington Post publishes story on how GSA continues to foil research by Blakey and colleagues by short-changing their requests for funds. Blakey disparages inadequate funding, telling *Post* reporter, "We don't need . . . 'a colored grant.'"

DAVID ZIMMERMAN

September 2002
Howard University, dismayed, asks to withdraw from ABG project. GSA looks for new sponsor but warns Howard of its liability if it fails to finish contractually agreed-upon work.

October 2002
Howard decides to stay in. It sets up new administrative processes to carry work forward with Army Corps of Engineers (ACE). Descendant Community in New York blames GSA for delays.

January 31, 2003
GSA and Howard announce they will choose peer reviewers for Final Reports.

February 20, 2003
GSA and ACE promise early completion of research and reburial of remains. GSA confirms that Blakey is no longer in charge; Trimble is. Trimble declines interview requests.

August 2003
Blakey and colleagues submit new draft of Final Report to internal reviewers, who reject it.

September 30, 2003
Human remains returned to New York City and reburied with much fanfare in ABG. Report deadlines continue to slip.

December 2004
Final Report submitted to GSA.

Early 2005
GSA accepts *Final Report* and releases it to public.

SANKOFA?

ACKNOWLEDGMENTS

This book has been a long journey, and I am deeply grateful to the many who have helped me along the way. David Baltimore started it when, in 2001, he sponsored, and invited me to attend, a press seminar on exemplary research being conducted by black scientists. One of them brought me up to date on a not-so-exemplary project that he had worked with briefly. It sounded newsworthy — and so, set me on my way.

The principals and the government sponsors of the African Burial Ground (ABG) Project resisted my inquiries.

My colleague and friend, Tom Watkins, pointed me to a key connection: Mary Cassandra Hill, of Tuscaloosa, Alabama, the ABG project's one roll-up-your-sleeves doctoral researcher. She had been fired when her work was almost completed. Fortunately, she saved the paper trail of these experiences — which she kindly shared with me.

Special thanks to Cassandra Hill! Special thanks to Jerry Rose!

A second "Cassandra," this one a federal public relations consultant to the project, also had strong memories of what had gone right — and what wrong: Cassandra Henderson explained these to me in the context of the project's political fortunes.

Speaking of PR, GSA's public relations official on the project, Renée Miscione, was quite helpful; she told no secrets. I have high regard for her and her work.

Conservator Gary McGowan pointed me to key sources. Howard Dodson authorized my access to the Schomburg Center's still-unsorted archive on the ABG.

This book was approaching completion in 2008 when the economy, and particularly the publishing business, stumbled toward failure. My story on the fate of long-buried black human remains became a hard sell. A number of science-writer colleagues and friends offered advice and support. They include Deborah Blum, Bob Calvin, Claudia Dreifus, Anne Moffat, Janice Tanne, and Ruth Winter. Computer doctor Jesse Larocque provided technical support. Anita Tester kept order. Ms. Stubby and her pride helped, too.

Marcia Newfield and Deborah Warner kindly read early drafts. Amanda Sutphin checked my account of efforts by her New York City agency.

Cynthia Barber was a patient and careful editor. Barry Sheinkopf carefully designed and laid out this book.

Three wonderful women worked with me — and so made this report possible. Wendy Gessner started out as typist and organizer. Alas, Wendy died. She was followed here by Cindy Ely, highly skilled, whose two young daughters were always a stitch to be with. When she moved away, Michelle Mitchell, a wonderful person and a professional, took over the keyboarding and stayed to the end. I thank them and all others who have helped.

I alone am responsible for the errors.

—David R. Zimmerman

I am grateful to the following for permission to reprint these images:
Jean Allen: front cover [Mary Hill]

Bill Denison: p. 167

Government Services Administration: front cover [Michael Blakey], p. 21, p. 35, p. 89, p. 101, p. 126, p. 130, p. 258 [bottom]

Mary Hill, p. 112, p. 131, p. 151

National Park Service: p. 31

Barry Sheinkopf, p. 4

John J. Spaulding, Friends of the Office of State [Connecticut] Archeology: p. 258 [top]

SANKOFA?

INDEX

Activity Report (See also *Quarterly Report*).................207, 212, 224, 227
Admiral Family Circle Islamic Community..65, 167
Advisory Council on Historic Preservation (ACHP)......29, 58, 75, 204, 213
African Burial Ground (ABG).........1, 5, 17, 11, 21, 28, 33, 35, 40, 51, 63, 76, 88, 99, 116, 127n, 134, 146, 174, 186, 232, 243, 253
    American Revolution...20, 127n
    Documentary (African Burial Grounds: An American Discovery)..83
    General Audience Report..260
    Integrated Report (See *Skeletal Biology Report*)
    National Historic Landmark............................21, 40, 56, 85, 228
    Project................12, 15, 26, 41, 116, 120, 129, 161, 197, 216, 224, 248, 256,
    Research Design....37, 41, 46, 58-63, 72-78, 88, 99, 103, 127n, 154, 173, 195, 238
    Steering Committee, Federal......................52, 54, 63, 77, 98, 118, 133
    Van Dusen Family...19n
American Association of Physical Anthropologists (AAPA)..139, 147, 244
Armed Forces Institute of Pathology (AFIP)..147
Army Corps of Engineers..206, 214, 225
    Center for the Curation and Management of Archeological Collections...206
Artex Fine Arts Services...98
Artifacts and Evidence (*See also* Health Evidence)
    Beads...124, 126-128, 128n
    DNA data............88, 93, 96, 104, 117, 125, 139, 173, 174n, 187, 192, 219, 226, 233
    Musket Ball...126, 134, 147, 235
    Shrouds...25, 124, 127n, 171
    Soil Pedestals (grave site)..40, 93, 124, 131
    Tooth Filing...127, 127n
Association of Black Anthropologists..46
Blakey, Dr. Michael L..40
    Belize..43
    Brent Study..81

Brown University..212
College of William and Mary..13, 212
Dissertation...78-82, 255
Mayan Villages...43
Messiah..92, 253
Smithsonian Institution...41, 42
York College..139
Cantwell, Anne-Marie.....................................18, 20, 23, 27, 34, 196
    Rutgers University...18, 68, 222n
    Unearthing Gotham, the Archeology of New York City...........17, 196
CNN...191
Cost of Project...............................2, 30, 99, 117, 138, 171, 174, 184-185, 261
Cultural Resource Management (CRM)...32
    Contract Archeology..31, 204
DC Employee Consultation and Consulting Service...................149
Descendant Community................2, 8, 14, 40, 55, 66, 74, 92, 156, 190, 201, 206, 214, 225, 234, 237, 248, 251, 256
Dodson, Howard..52-55, 77, 87, 90-92, 154, 214, 229
Dutch West Indian Company...146
Environmental Impact Statement..33
Foley Square..52
    Foley Square Project..24
    Ted Weiss Federal Building...17
Freedman's Cemetery, Dallas Texas..................................135, 184, 185
Freedom of Information Act (FOIA)........................15, 141, 179, 203
General Services Administration...............3, 29, 39, 51, 63, 75, 87, 98, 133, 138, 152, 168, 178, 205, 210, 224, 245, 257
    Final Report (See *Skeletal Biology Report*)
    Research Design....37, 41, 46, 58-63, 72-78, 88, 99, 103, 127n, 154, 173, 195, 238
    Research Design Panel..................92, 98, 106, 116, 118, 244
    Sneed, Peter............................11, 15, 84, 153, 181, 200, 213, 225
Health Artifacts
    Anemias...107, 109, 119, 122, 163
    Hyperostosis...111, 163
    Hypoplasias...107, 111, 126, 237
    Bone fractures..107, 147, 176, 230
    Porotic hyperostosis..111, 122, 130, 176
Health Index..113, 115
    Mark I Database..113, 240, 247
Hill, Mary Cassandra....112, 120, 127n, 142, 152, 171, 182, 208, 231, 244, 260, 262
    Alabama Indian Mounds..112, 120
    University of Alabama, Tuscaloosa..................................122

SANKOFA?

University of Tennessee, Knoxville..................................145
Historic Conservation and Interpretation, Inc. (HCI)..22, 37, 46, 58, 82, 91
    Rutsch, Edward S..................................22, 32, 37, 46, 91
Holocaust (Museum)..................................55, 68, 175, 194
Howard University..................2, 8, 41, 47, 59, 66, 82, 100, 118, 138, 169, 183, 191, 200, 202, 212, 232, 241, 245, 260
    Blakey, Dr. Michael L. (See Blakey, Dr. Michael L.)
    Cole, Dr. O. Jackson..................................213
    Department of Sociology and Anthropology..................................160
    Donaldson, Dr. James A..................................213
    Kittles, Dr. Rick..................................169, 186-197, 216, 233, 244
    W. Montague Cobb Biological Anthropology Laboratory....8, 45, 116, 132, 169, 182, 215
Jackson, Dr. Fatimah L.C..................................193, 226
John Milner Associates..................11n, 15, 32, 48, 59, 63, 85, 100, 133, 185, 204
LaRoche, Dr. Cheryl..................................15, 73, 127, 153, 156
Lehman College (See Metropolitan Forensic Anthropology Team)
Library of Congress..................................140
Mack, Mark E..................................15, 119, 124-137, 209, 239
Massachusetts Bay Community College..................................189
Metropolitan Forensic Anthropology Team........23, 33, 37, 40, 47, 53, 63, 82, 102, 154
    Eisenberg, Dr. Leslie E..................................24, 39, 46, 102, 148, 154
    Lehman College..................................24, 48, 57, 82, 100, 123
    Turkel, Dr. Spencer..................................24, 38, 47, 57, 64, 82, 156
Middle Passage..................................89, 127, 174n, 189
Moton Conference Center..................................210
National Historic Preservation Act of 1966 (NHPA)..........9, 22, 28, 56, 87, 104, 171, 249
    National Historic Preservation Program (NHPP)..................28
    Native American Graves Protection and Repatriation Act of 1990..................................29
    Section 106..................................30
National Human Genome Research Institute (NHGRI)..................186, 193
    DNA data..................................173n
    National Human Genome Centers..................................186
National Park Service (NPS)..................................30, 76, 155, 206, 228
Native Americans..................18, 42, 67, 104, 107, 114, 120, 141, 234, 260
Negroes Burial Ground (See African Burial Ground)
New School for Social Research, New York (NSSR)..................................85
New York City Commons..................................17, 20
New York Historical Society..................................240, 241
    Slave Conspiracy of 1741..................................148

Slave Insurrection of 1712..............................148
Slavery in New York..............................240
New York Landmarks Preservation Commission..........8, 38, 55, 56, 75, 206
New York Public Library..............................52, 140
    Schomburg Center for Research in Black Culture....14, 52, 154, 214
New York State Historical Preservation Organization (NYS HPO)....30, 33
Office of Public Education and Interpretation (OPEI)..............85, 128, 132, 168, 182, 201, 203, 230, 256
    Wilson, Dr. Sherrill D. 85, 128, 133, 182, 203, 256
Paleopathology Association..............................129
Peer-Review..............88, 92, 96, 99, 194, 202, 208, 209, 224
Powderhorn/Phillips Cultural Wellness Center..............................218
Quarterly Report (See *Activity Report*)
Race/Racism
    Biological..............................43, 53, 67, 69, 221, 222
    Black..............................55, 84, 154
    Colored Grant..............................211
    Darwin, Charles..............................222
    Traditional..............2, 8, 13, 36, 43, 58, 68, 73, 79, 82, 85, 146, 154, 165, 198, 211, 214, 22, 222, 240, 247, 253
Rankin-Hill, Lesley..............................15, 59, 106, 109, 119, 230, 235, 243
    First African Baptist Church Cemetery..............48, 59, 106
Rites of Ancestral Return..............................227
Rutgers University..............................18, 68, 222n
Sankofa..............................5, 130, 229, 258, 259
    Adinkra Cloth..............................6, 258, 259
Savage, Augustus (See also Committee on Public Works)..........49, 57, 261
Schomburg, Arturo..............................154
    Vindicationism..............................153-156, 251
Science Citation Index..............................245
Scientific procedures
    Coding of Data..............................131, 163, 164
    Coroner's Methods..............................26, 27
    DNA Analysis..............................174, 180
    Multivariate Discriminant Function Analysis..............69-71
    Paleo-demography..............................235
Skeletal Biology Report
    ABG History Final Report..............................21, 231, 260
    First Draft..............................168, 175, 178, 195, 229
    GSA Response..............................178-179
    Skeletal Biology Review Board..............224, 227, 229, 230
Sneed, Peter..............................11, 15, 84, 153, 181, 200, 213, 225

SANKOFA?

State University of New York, Buffalo..................................155, 257
Statistical Research, Inc. 260
Trimble, Dr. Michael K. Army Corps of Engineers...........206-208, 214, 226
Trinity Church Cemetery..................................18, 67, 138, 234
United Nations Educational, Scientific, and Cultural Organization
    (UNESCO)..................................68, 222
University of Arkansas, Fayetteville..................................72
    Cedar Grove Cemetery..................................88, 108, 119, 184, 236
    Rose, Dr. Jerry C. 72, 88, 97, 106, 108, 119, 151, 164, 178, 184,
        205, 234, 247, 249
University of Maryland, College Park..................................213, 226
University of Massachusetts, Amherst..................10, 43, 73, 111, 119, 145,
    163, 182, 226, 255
    Anthropology Department..................................44, 119
    Anthropological Department Admissions Guide..................78
University of Tennessee, Knoxville..................................145
    Forensic Anthropology..................................145
Update..................85, 124, 128n, 132, 166, 167, 172, 173, 175, 182, 183, 193-196,
    201, 256
US Congress..................................17, 29, 49, 52, 63, 98, 140, 200, 201, 249
    Committee on Public Works..................................49
US Department of the Treasury..................................117
Wall Street..................................34
    Governor's Island..................................18n
    Slave Market..................................35, 101
Wall, Dr. Diana diZerega..................................18, 20, 23, 27, 34, 196
    City College of New York..................................18
    Unearthing Gotham, the Archeology of New York City............17, 196
White House Office of Budget and Management (OBM)..................191
Wilson, Dr. Sherill D. (See Office of Public Education and
    InterpretationandUpdate)
World Trade Center..................................24, 85, 168
    6 World Trade Center..................................85, 183, 202
    9/11..................................24, 202

David Zimmerman

DOCUMENTS/PUBLICATIONS INDEX

Activity Report (See also Quarterly Report)..........207, 212, 224, 227
American Journal of Physical Anthropology..........46
Another Dimension to the Black Diaspora: Diet, Disease, and Racism..........109
Archeology Final Report..........89
Boston Globe..........188, 192
Breaking Ground, Breaking Silence: the Story of New York's African Burial Ground..........25, 195
Chronicle of Higher Education..........55n
Current Biography Yearbook..........42, 157
Death in the New World..........257
Gone to a Better Land..........108, 184
Historical Archaeology..........153
International Journal of Anthropology..........46
Los Angeles Times..........193
National Geographic..........167
Nature..........243
New York Daily News..........198, 256
New York Post..........47
New York Times..........11, 27, 151, 192, 216, 241, 256
Newsday..........202, 256
Quarterly Report (See Activity Report)
Roots..........34, 36, 188
Skeletal Biology Report
    ABG History Final Report..........21, 231, 260
    First Draft..........168, 175, 178, 195, 229
    GSA Response..........178-179
    Skeletal Biology Review Board..........224, 227, 229, 230
The Backbone of History: Health and Nutrition in the Western Hemisphere..114
The Communist Manifesto..........80
The Crisis of the Negro Intellectual..........92
The Descent of Man..........222
The Peculiar Institution: Slavery in the Ante-Bellum South..........108
Time on the Cross: The Economics of American Negro Slavery..........109

SANKOFA?

Update..................85, 124, 128n, 132, 166, 167, 172, 173, 175, 182, 183, 193-196, 201, 256
USA Today..241
Unearthing Gotham, the Archeology of New York City............................17, 196
Village Voice..57
Washington Post...210

Made in the USA
Charleston, SC
28 October 2013